BOYD'S MEAL

OF NEW ENGLAND

Their Genius, Madness, History & Future

by
Daniel Lombardo

Cape Cod, Massachusetts

Edited by Stuard Derrick & Adam Gamble.

Production supervised by Adam Gamble.

Book design, cover design and typesetting by Joe Gallante,
Coy's Brook Studio, Harwich, MA 02645.
For more information on book design by Coy's Brook Studio
e-mail: coysbrookstudio@comcast.net.

ISBN: 0-9719547-7-1

For more information about
New England Windmills
please contact:
On Cape Publications
P.O. Box 218
Yarmouth Port, MA 02675
Toll free: 1-877-662-5839
On the web at www.oncapepublications.com
email: windmills@oncapepublications.com

First edition
10 9 8 7 6 5 4 3 2 1

Printed in Korea.

Table of Contents

Introduction

The Genius and Madness of This Curious Machine

"He it was who climbed the slender latticed arms and set the sails; he it was who hitched the oxen to the little wheel to turn the white wings into the wind's eye; he it was who touched the magic lever and presto! The long wings beat the air, the great shaft began to creak and turn, cog played on its fellow cog, and the mammoth stones began to revolve...."

Shebnah Rich, *Truro – Cape Cod*

In the 1960s, I set up a tent in the pine woods of Cape Cod and kept a journal. I suppose I went to that landscape looking for insight and inspiration, propelled by Henry David Thoreau and his vivid Cape Cod journals of a century before. From then on, my imagination has always held images of clear blue oceans, wooden sailboats and their twins, those curious, creaking windmills occasionally found onshore like abandoned ships.

Naturally, a boy's dreams are soon tested. Having pitched my tent at night under the stars, I discovered the next morning that I had done so in a luxurious patch of poison ivy. Itchy but undaunted, I went on. At the beginning of the 21st century, I'm returning to old scenes, and exploring new regions of New England in search of windmills. Older now, I'm finding much more than I bargained for.

The romance of the old windmills still survives. It's being kept alive by people like white-bearded Jim Owens, who runs the oldest of Cape Cod's remaining windmills in Eastham, and by Patrick Prugh and the other young apprentice wind millers at the Old Mill on Nantucket Island. The old stories of children snagging rides on windmill blades, of millers breaking bones, or cattle losing their lives in those same blades, are still told. Follow the Windmill Trails I've created for this book, and you'll find a lost world of New England villages that revolved around windmills—where getting one's "daily bread" was a communal activity, and the mill was as important to its community as the Town Hall or church.

An early postcard of the oldest remaining windmill on Cape Cod, Eastham, MA.

In search of the windmill's place in the new century, I've discovered a new breed of devotees. Scattered around New England are people like artist Samuel Barber, guitar maker Steve Connor, and flight instructor John Falvey, who recycle old mills in new ways.

Artist Samuel Barber and his reproduction windmill in Hyannis Port, Cape Cod. Photo by the author.

I've met 21st century windmill keepers like Art Miller, whose product is not cornmeal but electricity. Wind power is, in fact, the fastest growing source of electric energy worldwide—and it's based on a machine that goes back more than two-thousand years.

The Searsburg Wind Power Facility, 2003. Photo by the author.

My innocent image of the windmill as a symbol of pastoral bounty has been deepened by this journey. I've come to recognize what writers like Cervantes, Tolstoy, Dickens, Poe, and Sylvia Plath have seen in windmills. I've looked again at Don Quixote and his "terrifying and never-before-imagined adventure of the windmills," and at the artists and musicians who have found profound meaning in them. The playful images of windmill sails turning lazily by the sea often shade into emblems of destruction and madness. Yet, as Cervantes said, "everything turns in circles." A young Hong Kong pop group recently named itself "Windmills" to evoke wildflowers, rolling fields, a peaceful heart, and a clear mind.

Windmills have always had a mysterious power to capture the imagination. They began as practical tools of survival, providing the bread of life, water for fields, and salt for the table, but quickly became mythic beings. No one knows when or where windmills originated. Some scholars believe that Babylonian leader Hammurabi irrigated crops with wind power in the 18th century B.C., but no evidence survives. Many European folktales refer, tantalizingly, to windmills that, if they existed, left no trace. For example, the 3rd century Irish king, Cormac Ilfada, it was said, fell in love with the Princess Ciarnute. When his queen found out, she condemned the princess to the endless task of grinding corn by hand. The princess begged mercy from her lover, the king, and he sent for a Scottish millwright to build a windmill to grind the corn for her.

Windmills, it was believed, captured the wind that was sent by the gods, just as water mills partook of the power gods sent through water. In many cultures, the transition to water and wind power produced a sense of spiritual and economic discomfort. In ancient Rome, for example, mills were at first powered by donkeys and slaves. The first known reference to a water mill in the Western world came in 85 B.C., when Antipater of Thessalonica wrote, "Cease your work, ye who labored at the mill. Sleep now, and let the birds sing to the blood red dawn. Ceres has commanded the water nymphs to perform your task; and these, obedient to her call, throw themselves on the wheel, force round the axle tree and so the heavy mill."[1] Around the same time, water mills began to appear in China; at the palace of Mithridates, King of Pontus (northeastern Asia Minor, along the south coast of the Black Sea); and in Jutland (present-day Denmark and West Germany).

But the Roman Empire had several issues with water mills: First, if the goddess Ceres commanded the water nymphs of every river and waterfall, perhaps there would be unpleasant ramifications if humankind harnessed her power. Second, the empire employed huge numbers of slaves in the mills to provide food for the masses. For over 300 years, thousands of Roman citizens were, basically, on the bread dole. At first, bushels of grain were distributed; then, as Edward Gibbon described in *The Decline and Fall of the Roman Empire*:

> *For the convenience of the lazy plebeians the monthly distributions of corn had been converted into a daily allowance of bread; a great number of ovens were constructed and maintained at the public expense; and, at the appointed hour, each citizen who was furnished with a ticket, ascended the flight of steps which had been assigned to his peculiar quarter or division of the city, and received...the weight of three lbs. for the use of his family.*[2]

If water mills replaced slaves, thousands would have been thrown out of work, which would have resulted in slave revolts that might have brought down the empire.

Nevertheless, the Roman Empire fell on its own, and waterpower replaced slave and animal power. Windmills, on the other hand, would never replace water mills wherever a good supply of flowing water was available. In low, flat sandy areas, streams provided barely enough power to turn a child's pinwheel, or were affected by ocean tides. Thus, in coastal Europe and places like Cape Cod and the coast of Rhode Island, windmills proliferated. Water mills, unlike windmills, were also subject to drought, flood, winter freezes, and rot from continued soaking in water. Windmills had their own problems, including, as we shall see, gales, hurricanes, tornadoes, fires, and individuals who

The Truro windmill in ruins, with the Highland Lighthouse in the background. Photo courtesy of the Cape Cod National Seashore Archives.

tended to walk into their swirling blades. Historic windmills have also had to suffer the humiliation of being shunted aside by historic lighthouses that, through their nagging threat to topple into the sea, have taken the lion's share of preservation funding.

The world's first documented windmills date from 644 A.D., in Seistan, Persia, now Iran. By the 10th century, Arab historian Masudi wrote of mills that were already very old and had been introduced to Persia by an Egyptian warrior in the time of Moses. In luxurious Persian gardens, he wrote, "...wind turns mills which pump water from wells to irrigate the gardens. There is no place on earth where people make more use of the wind." [3]

By the 13th century, horizontal windmills were pumping water, grinding corn, and redistributing drifting desert sands. Historian John Reynolds described the European origins of this unwieldy contraption: "The typical European windmill was almost certainly an independent invention of the Gothic North. In the twelfth century this curious machine, perched on its single leg like some monstrous sciapod from the pages of a bestiary, must surely have seemed a work of genius bordering on madness." Reynolds thought it possible that the first of the Northern European windmills was built in England, in the little village of Weedley in Yorkshire, and notes documentary evidence at a date of 1185 A.D.[4]

Woodcut of 17th century Dutch post mill.

Historians agree that the windmills of Persia and China were made of horizontal paddles set up somewhat like a carousel. But how did vertical blades come to be perched on towers in Western Europe? Some scholars, like Suzanne Beedell and Volta Torrey, skirt the issue and merely state that horizontal mills originated in Eastern countries and vertical ones in the West. John Reynolds developed the theory that the East and West came up with windmills independently. He further posits that, in a reversal of the prevailing flow of ideas from East to West, the more efficient vertical mill made its way from Europe to the Eastern Mediterranean region and then the Far East.

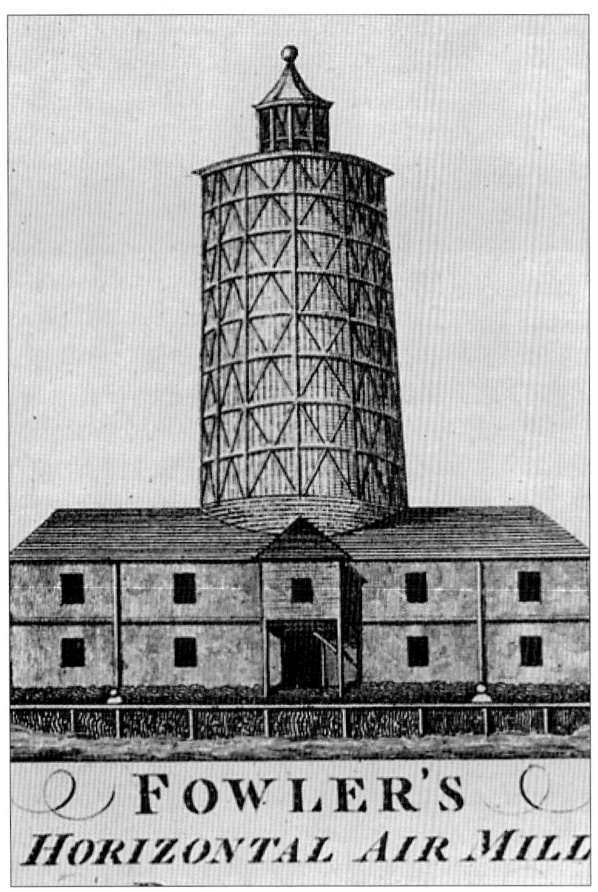

An English 19th century horizontal windmill. From C.K Skilton's *British Windmills and Watermills*, London: Collins, 1947.

It wasn't until the 17th century that the windmill would make its first appearance in North America. It has been said that the first concern of a new community in America was to build a meeting house and a mill, and there is some truth to this. Shebnah Rich, for example, wrote in his history of Truro, Massachusetts, that during the town's early years "the principal landmarks seen upon approaching Truro shore from the bay side, were the tall grist mill, and the two meeting-houses, grouped but a few hundred rods apart, in a triangular position."[5] It would be another thirty years before Truro built a town hall.

The most important elements of the diet the English settlers brought with them to the New World were meat, fish, bread or porridge, and beer. According to James Baker, historian at Plimoth Plantation in Massachusetts, field crops in the earliest days would have included winter and summer wheat, rye, meslin (wheat and rye sown together), barley, and corn. Corn was the most important crop and was processed by soaking in water, then pounding to remove the hulls, which were then winnowed away. The grain was ground in either a mortar, a water mill, or a windmill, after which it was sifted. The result was either a fine flour or a coarser meal called "samp."

Cereal grains were also used for brewing. Barley was the most popular grain to be malted, though there are references to Indian corn malt. Malted grains were brewed with hops to produce beer, or without hops to produce a sweeter ale. Many households produced two sorts of beer, a "small" beer with a low alcohol content, that was consumed by every member of the family, and a "strong" aged beer.

Historian Frederick Freeman leaves out this particular use of grain in his account of the typical colonial Cape Cod diet:

The meals, in those days, were frugal, the course at dinner being, in winter,

> *ordinarily, first, "porridge"– a broth, with a few beans thrown in, and seasoned; second, an Indian pudding, with sauce; and third, boiled pork and beef, with potatoes and pumpkin. Suppers and breakfasts were usually alike—milk with toasted bread in it, or sweetened cider, hot in winter, with bread and cheese. On "Sabbath mornings," they indulged in chocolate or tea, the first sweetened with molasses, the tea with brown sugar, and the concomitants were pancakes, dough-nuts, brown toast, or pie. They had no Sunday dinners until 'both meetings' were over—but then the intermission was short; after meetings, a spare-rib, a stew-pie or roast beef, goose, chicken or turkey, made up the repast, with a few et ceteras. In spring, summer, and autumn, bread and milk constituted the principal break-fast and supper. The chief exception to the above bill of fare was fish, which in its varieties was abundant.*[6]

The first permanent English settlement in America was in Jamestown, Virginia, in 1607. Unlike later colonists, the earliest settlers, like those at both Jamestown and Plimoth, hadn't the time nor the expertise to build mills in their early years. Both colonies relied at first on flour and provisions sent from England. Jamestown successfully grew a cash crop of tobacco to send back to its investors, the Virginia Company of London, and only later turned to growing corn and wheat; by 1619, the colony was growing all three crops. Meanwhile, the local Native Americans ground their grain with mortars and pestles, and possibly with common hand querns.

By 1621 there were some 1,500 people in Virginia to feed. At a meeting in London of the Virginia Company on July 7, 1620, the secretary recorded, "Itt is verie necessaarie for the benefitt of the Colony that divers skillful Millwrights be provided and sent to sett vpp Corne watermilles in the seuerall parts of the Collony." A year later, the company ordered "...that yor corne Mills bee presentlie Erected and puplique Bakehowses in eurie Burrough bee built with all speed and dilligence." And they were, for the Records of the Virginia Company refer in January of 1622 to "the good Example of Sr. George Yardley by whom a wind mill hath beene allredy built, and of mr. Thresurer who ys about the Erecttinge of a water mill, (which) wee hope be greate encouragements to others in a matter of soe greate and generall use."[7]

America's first windmill—in fact, the first mechanical mill of any kind—was built in the Virginia Colony in 1621. It was erected at Flowerdew Hundred, the private plantation of Sir George Yeardley, Governor and Captain-General of Virginia. The French began to build windmills in North America a few years later. In 1629, they constructed one on the banks of the St. Lawrence River, and followed it with others between Lake Erie and Lake Huron. French windmills also went up near St. Louis on the Mississippi River in the 1700s, and at Detroit. The Dutch, of course, brought a windmill tradition with them, as did the Swedes, Germans, and Portuguese. A famous Swedish windmill was put up at Lawrence, Kansas, in the 1800s, and "Bluff's

Windmill" ground corn in Washington, D.C., where the Kennedy Center for the Performing Arts is today.

In America we tend to forget how common windmills once were in our coastal communities. We see evidence of water mills, with their stone dams and sluiceways, throughout the countryside; but windmills, made of wood and useful only in specific, limited landscapes, disappeared all too easily.

Few people realize that New York's skyline, the most famous in the world, was once graced by windmills. Before Broadway became famous for its uptown theatres, it was known for the windmills that lined its lower reaches. The first windmill built for the use of the new Dutch town on Manhattan Island was located "near Broadway, between...Liberty and Cortlandt streets," in other words, on the location where the World Trade Center towers would stand over three-hundred years later.

According to an article in *Scientific American* published in 1867:

> *Minuit, the first Dutch governor, built, according to Moulton, "two or three wind mills at Manhattan, by which corn was ground and boards sawed." One of these, a flour mill, stood on a hill which occupied a part of the present Battery, so near the fort that the latter, which was rebuilt...in 1633, intercepted the southeast wind, and rendered the mill nearly useless. But one of three wind mills previously erected was in operation in 1638....*
>
> *On one of their farms, of which they reserved several in different parts of the island, the West India Company erected a "Wint molen"* [wind mill] *for the use of the town. It stood near Broadway, between the present Liberty and Cortlandt streets...."Old Wind Mill Lane," running from Broadway to Greenwich street, and between Cortlandt and Liberty street...was, in 1729, the most northern street west of Broadway.*
>
> *Mills of this class were also built by private enterprise. Jan Teunizen had a wind mill in 1665... near the corner of Chatham and Duane streets.... A wind mill once stood on the hill in the rear of the old jail, or the present Hall of Records, and an eminence near the Chatham Theater was called "Wind Mill Hill."*[8]

In 1684, New York was given the exclusive right of bolting flour within the colony and even more mills were added to the island. Two years later, the city received a new charter and a new seal. This seal, reflecting the city's principle sources of prosperity, retained the beaver from its earlier seal and added a windmill and flour barrel.

New York City was reminded of this part of its past, surprisingly, in the aftermath of the horrific terrorist attacks of September 11, 2001. Nine months later, while committees debated a memorial to place on the site of the World Trade Center, *The New York Times* published a remarkable article by John Tierney entitled, "In New York, Change Is Traditional." Tierney laments New York's history of razing the old in order to build the profitable:

Who even remembers that the city's skyline and economy were dominated in the 17th century by another structure at the World Trade Center site, a windmill? Winners may write history, but they also build over it.... The marketplace may at last be telling New York to start preserving its past, and there's no reason we have to limit ourselves to the history of our worst day. It might even make sense to put that windmill back up.[9]

At the heart of any windmill was the grindstone, and its mythic place in America is no better evident than at the site of the Eternal Flame at the John F. Kennedy Memorial in Washington. For several generations the Kennedys have been associated with Hyannis Port, on Cape Cod. J.F.K. carved his initials on the beam of a windmill there as a boy. Following the assassination of President Kennedy on November 22, 1963, Jacqueline Kennedy commissioned John Carl Warnecke of Washington, D.C., to design the memorial that was to become one of the nation's most revered places of pilgrimage.

Mrs. Kennedy wanted to tie the resting place to Cape Cod, so she requested that the memorial be paved with granite from the Cape. The John F. Kennedy Memorial's most important element, an eternal flame, was still to be designed however. It was antique dealer Orville Garland who suggested a local millstone to Mrs. Kennedy. Florence Tinkham had a millstone in her Woods Hole garden, and she had died just the previous year. While razing the house, crane operator Richard Fish noticed the five-foot-in-diameter millstone in her backyard. Overgrown by shrubbery, it had once been the center of a radial garden. Mrs. Kennedy saw immediately the exquisite and fitting idea of this pink granite millstone, both as a design element and a symbol. Hundreds of thousands of visitors make their way to the Kennedy Memorial each year and gaze at the eternal flame held by this Cape Cod millstone, a symbol of America and J.F.K.'s place in its heart.

A few fortunate people, like the artist, the guitar maker, and the flight instructor you'll meet in this book, live in windmills. Unless kept as museum pieces, or incorporated into homes, however, the vast majority of windmills left little but their elegantly carved stones.

Most New Englanders think of windmills only as images on Dutch tiles, unless they frequent the windmill meccas of coastal Rhode Island and Cape Cod. Perhaps they remember when every miniature golf course had a windmill, and one had to time a shot to a precise moment between sweeps of its arms. Many garden centers stock windmill lawn ornaments which offer a picturesque touch to a yard, but few connect these with the organic splendor of the real thing. Everyone seems to know that Don Quixote, in Cervante's masterpiece, "tilted at windmills," but they aren't certain why. And when asked if they've seen a windmill lately, many tourists will refer you to a lighthouse.

In this book, my deepest hope is to revive the windmill as a living, churning, creaking part of both the physical landscape and of our imagination. I want us to place the sleek new windmills, generating energy on hilltops and offshore, within the long sweep of windmill history. We'll look to and beyond the windmill's critical role in early history: the brief niche in time when they provided bread and salt; pumped water to crops, or out of wetland in places like the Netherlands; pumped water for homes and shops; and even ground bark for tannin, and chocolate for drinks. In traveling through New England, suddenly place names like Windmill Hill in Litchfield, and Windmill Point in Stonington, both in Connecticut, or Windmill Cottage in East Greenwich, Rhode Island, will make us look twice. We can imagine a landscape when nothing but the wind, whether sent by the gods or from the prevailing westerlies, was needed to power everyday life. A time when windmills were integrated seamlessly into our poetry, our music, and our myths. When they were fanciful symbols of imagination, images of Biblical dignity, Shakespearean revenge, and Dickensian gloom. Or when New England counterbalanced between Whittier's windmills as emblems of the picaresque New England ideal, and the windmills of Poe's fevered mind. ●

19th century photo of Wyatt's Windmill, Middletown,RI, with sails turning.
Photo courtesy of Middletown Historical Society.

Chapter One

New England's First Windmills: *The Sharp Teeth Biting the Corn*

"The windmill was brought downe to Boston, because (where it stoode neere Newtown) it would not grind but with a westerly winde."

Massachusetts Governor John Winthrop's diary, August 14, 1632

The windmill at Windmill Point, Boston, in 1833. *Old-Time New England*, Jan. 1931, p.98.

I was sitting in a little restaurant called Artu on Prince Street in the Italian North End of Boston. Through the front window I could see dozens of tourists passing by, some looking for Italian delicacies, a gelato, or a cannoli, others meandering the brick sidewalks up to Paul Revere's House, or to Old North Church.

As I waited for my plate of garlic Gorgonzola pasta to arrive from the kitchen, I looked at the back of the menu. Among the North End sites the menu encouraged me to see was Copp's Hill Burial Ground. Originally owned by cordwainer (shoemaker) William Copp, the land was bought by Boston in 1659 for a cemetery. The menu casually noted that the spot was also known as Windmill Hill.

After hurrying through my meal, I walked up to the corner of Snow Hill and Charter Streets. There, on a windy hill overlooking Boston Harbor, was the very site of New England's first windmill.

Boston had been known by the Algonquin Indians as Shawmut, which meant "living fountains." Several streams, which were later harnessed to power watermills, splayed out over the peninsula; but the land featured several small hills ideal for windmills, as well. The English at Charlestown called the area Tri-Mountain, because from there they could see Beacon Hill, Copp's Hill and Fort Hill. European settlement brought European diseases, and by the early 17th century much of the local Native American population was decimated.

In 1629, the Massachusetts Bay Company in London chose John Winthrop to govern its colony in New England. On March 30, 1630, Winthrop and about 700 Puritans sailed from Yarmouth, landing at Salem, Massachusetts, on June 12. Within the year, Winthrop led his Puritans from Salem to Boston and established it as the capital of the Massachusetts Bay Colony. Within a decade the English population there quickly outpaced the population of Plimoth colony to the south by tenfold. Shipments of grain from England couldn't keep up with the demand from the colonies, so early on mills had to be built. Corn, in fact, was so vital to colonial America that it was used as a universal currency for trade in the first half of the 17th century.

In 1632, the first windmill in New England was built in Massachusetts between Watertown and Cambridge (or, to add to the confusion, "it stoode neere Newtown," as Winthrop wrote in his diary). It ground corn there only briefly, for it worked very poorly, or not at all when the inland winds came from the west, as they often did. In August of that same year, the windmill was moved to the waterfront location in the north end of the Boston peninsula. The hill it was set upon was dubbed Windmill Hill, and its 50-foot eminence spread between Prince and Charter Streets. It was, however, high enough to catch the sea breeze, and was said to be well protected from "three great annoyances, of Wolves, Rattle-snakes and Musketos." Later it offered the British an opportune point from which to lob artillery shells on the Patriots, as the English Admiral Graves did with a battery of six guns and howitzers on June 17th, 1775. Graves launched bombshells that set fire to Charlestown, then ordered the marines from the frigate Somerset in to finish the job.

William Copp owned part of this hill from the beginning of the settlement. Not much is known about this shoemaker. When he wrote his will in 1662, he called himself "sick and weak," and he left "the enjoyment" of his house (but not the actual house), to his wife Gooddeth (Judith). To his daughter Ruth, he left "my great kettle, little pot and chaffen dish." His hill was for a time known as Snow Hill, because of its talent for attracting and retaining snow in the winter, but by the time of the revolution, it appeared as Copp's Hill on maps.

Boston's North End became the place where the town's well-to-do settled and

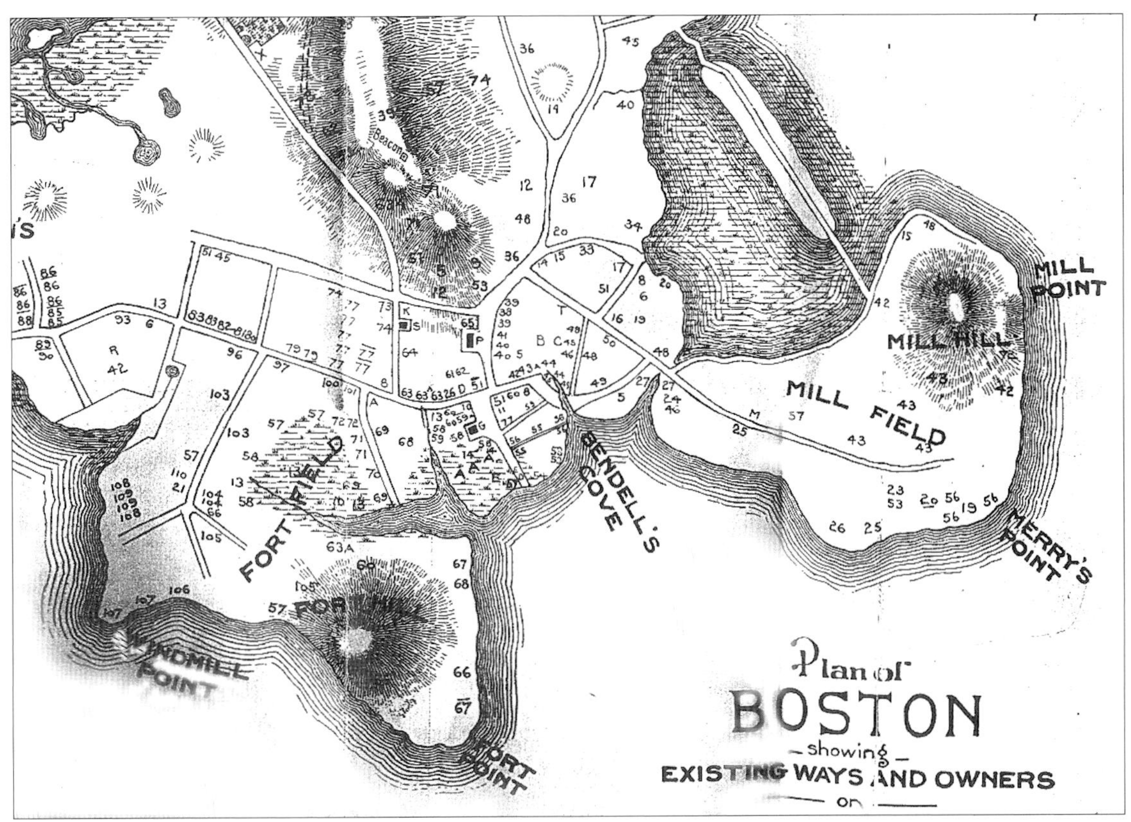

Map of Boston, 1635, with Mill Hill in the North End. Site of New England's first windmill. Note Windmill Point, lower left, site of many windmills. Map compiled by George Lamb, 1903.

continued as such until the colonial revolution. Thus, in 1659, when another cemetery was needed in addition to Boston's original King's Chapel Cemetery, Copp's Hill was purchased. In 1723, just yards east of Copp's Hill on Salem Street, Old North Church was built. Two lanterns were hung in its steeple to signal Paul Revere in Charlestown to begin his famous ride to Lexington in 1775.

In 1733, years after Copp's Hill had become a burial ground, a windmill still stood on the site. In that year, the windmill was struck by lightning, and shortly afterwards it was moved to Salem, where it gave good (we hope) service as late as 1771.

After the Revolutionary War, Boston's first substantial African-American population settled at the northeastern corner of Copp's Hill, and the neighborhood became known as New Guinea. The hill itself lost a height of seven feet during this period as it and Beacon Hill were shaved to fill in the Mill Pond for more housing. The cemetery itself had been deteriorating since the war, when the British used some of its gravestones for targets, mutilated others "whenever a patriotic epitaph thereon excited their ire or envy," and took many of the stones to build a hospital for their wounded. According to Thomas Bridgman's 1851 book, *Epitaphs from Copp's Hill Burial Ground, Boston*, "The worst enemy to the memory of the dead—disgraceful as it may appear—was among our own citizens.... Individuals carried [gravestones] off with impunity to cover drains, make foundations for chimneys, lay at the bottom of tombs for coffins to rest on, or at their mouths to close up the aperture."

I tried not to let any of this (and Bridgman's other, more grisly details)[1] disturb my pleasant stroll around Copp's Hill. The headstones of Increase, Cotton, and Samuel

Mather, along with several other Boston figures are still here. Some stones are set along curving paths, obviously no longer located over their intended subjects. And at the crest of the hill, very near the center, is one unmarked, unexplained millstone set in the grass, a mute reminder of the oldest windmill in New England.

The windmill on Copp's Hill had been noticed early on by one of the first travel-

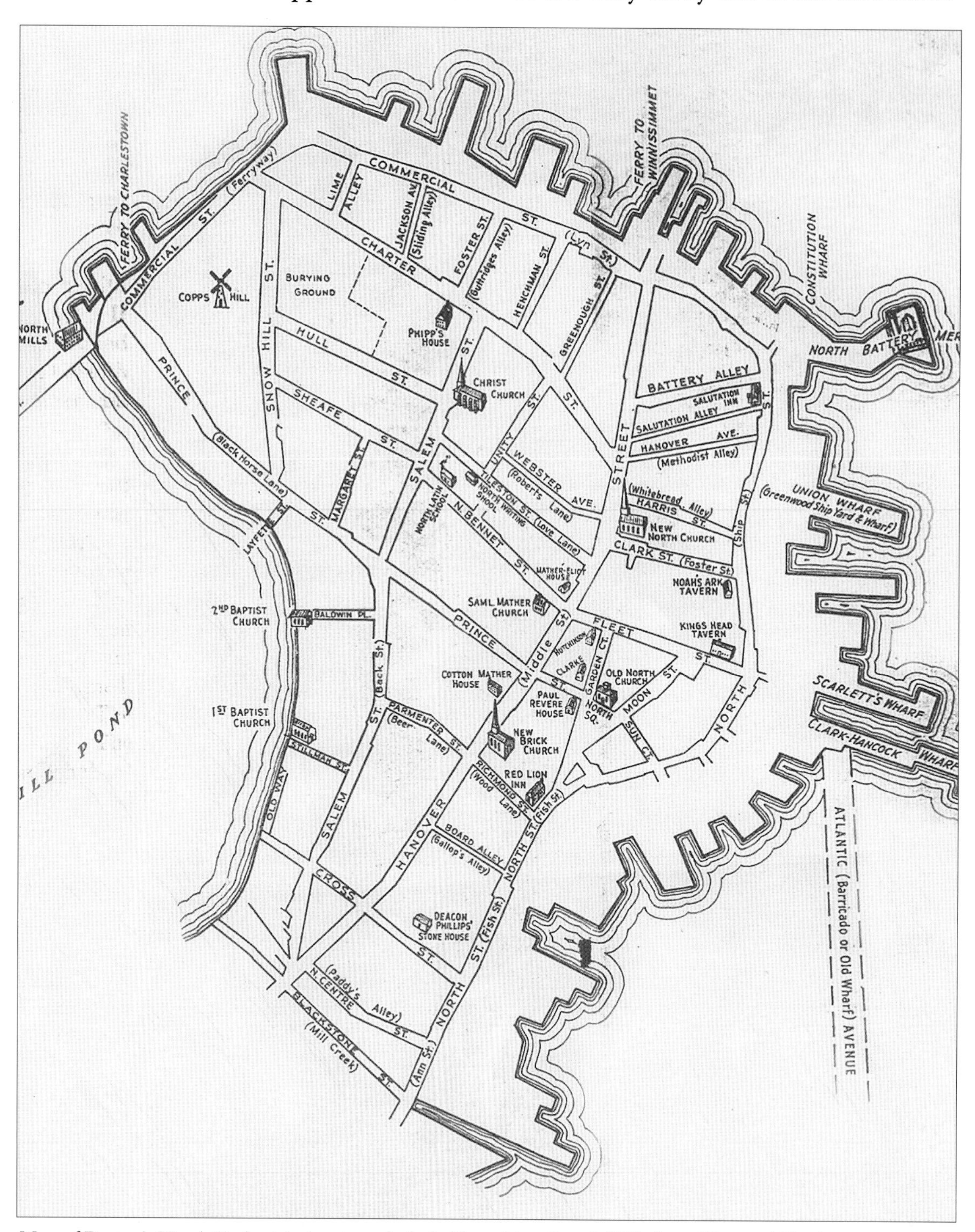

Map of Boston's North End, with drawing of windmill marking Copp's Hill, from *Historical Sketch...to the Copp's Hill Burial-Ground*, Boston, 1901.

ers to record impressions of Boston. William Wood's *New England Prospect*, published in London in 1634, records, "On the North side is another Hill, equall in bignesse (to Fort Hill), whereon stands a Winde-mill." Wood, of course, had seen many windmills in old England so he merely mentions in passing the Boston mill. He does happen to record, however, the vivid reaction of Native Americans to this peculiar new machine: "They do much extol and wonder at the English for their strange inventions, especially for a windmill which in their esteem was little less than the world's wonder, for the strangeness of his whisking motion and the sharp teeth biting the corn (as they term it) into such small pieces, they were loath at first to come near to his long arms, or to abide so tottering a tabernacle...."

In 1636, Winthrop's diary records a mill built on Windmill Point to the southwest, but on land that has long since been filled in and obliterated. Winthrop notes that three more windmills were put up before 1650. Nearly two centuries after the first windmill was erected in Boston, when the town officially became a city in 1822, there were still two windmills on Windmill Point. *Snow's History of Boston*, published in 1824, shows a large mill on Windmill Point, on the easterly side of the South Cove. A look through the official records of Boston reveals a surprising number of references to others:

> *This 31st of 7th moneth, 1642.*
> *There is liberty granted unto Widdow Tuthill to remoove her windmill into the fort, there to place it at the appointment of Capt. Gibones.*
>
> *This 27th of 12th mo., 1642.*
> *William Tynge, Capt. Gibones, and John Oliver...are appointed to view the swamp in the mill feild which Christopher Stanly desireth to buy of the Towne...and to lay out a convenient way through the milfeild, to the windmill and Charleton ferry.*
>
> *This 29th of 3d mo., 1643.*
> *Wm. Hibbins, Gent., and Wm. Colbron are appointed to lay out the High way to the South-Wind-milne that lyeth betweene Thomas Wheelar's and Robert Woodward's gardens.*
>
> *This 18th of 1st mo., 1644.*
> *There is graunted unto Wm. Teffe a parcell of land neere the south wind mill to be layd out by William Colbron & Jacob Eliot provied that He shall fence it with posts, and rayles, and not build upon it nor plant it with Indian corne nor any thing that may hinder the windmill.*
>
> *The 31st, 9th mo., 1649.*
> *It is ordered that the owners of the wind millne successively shall secure the milne from doing any damage to any Cattell or swine, upon penalty of making satisfaction sofficiently.*[2]

The landscape around Massachusetts Bay featured innumerable windmills as the population grew. John Humphrey settled in Lynn in 1634, and two years later he built one at a location later known as Windmill Hill. The miller there was Thomas Coldom, who had kept windmills in England before coming to the colonies.

Though windmills were largely privately owned, often financed by shares sold to investors, towns regulated them. They were, in effect, privately owned, publicly regulated utilities. Salem granted land for one in 1637, on or near the burial ground. In 1678, William Bowditch and a group of Salem investors formed a company to construct a windmill in Marblehead, on Rhodes Hill, and hired Newberry millwright John Woolcott to build it. It was to be round, 26 feet in diameter, and "to be built of good substantial timber, and completed to the turning of the key." Woolcott was to be paid eight score pounds, half in silver and half in fish or other goods. Unfortunately, the deal went wrong and all parties ended up in the courthouse in one of the earliest of windmill lawsuits.

In the lawsuit brought for poor workmanship, it was stated that the millstones "want floworing" and that "the incke and spindel were not sufficient neither for waight nor workemanship." Complaints were made about the "hopar and many of the utenssels of the mill," including the vanes, arms, sails, and "upper running girts." There was no brake to stop the mill (as essential as the parking brake on a car), and seven men could not turn her into the wind. One witness testified that Woolcott had said "that he would make the mill so good that she would perform her work with as little wind as a man might carry a candle burning open in the air without blowing it out, whereas, she would not work with a good gale of wind." [3] When William Bowditch died in 1681, he still owned five-eighths of this worthless contraption.

Miller Jonathan Wade kept a mill in Ipswich before 1667, located on a site called Windmill Hill. (Many inland New England towns had a Windmill Hill, just as coastal towns often had a site called Windmill Point.) Around 1680, a very fine millwright, Thomas Paine of Eastham, built a windmill in Plimoth. It still stands, capable of grinding corn to this day, in Eastham, where it migrated in 1793, after spending some years in Truro.

Newbury had a windmill as early as 1703, on a hill at the southern end of Frog Pond. It was used to grind corn and wheat, and inadvertently ground the head of a man who was examining the machinery in 1771. He died. The mill continued until 1774. Newbury also had a windmill for pumping water at its saltworks, set up soon after the Revolutionary War. Brockton built a windmill in 1781; and Swansea had one known as Potter's Mill, which still stood, though in ruins, in 1931.

Salem considered building a windmill to grind bark in the town's tanyards in 1793. (Ground bark was essential for tanning leather.) In 1794, Addison Richardson built a windmill on East Street in Salem, but he ran into problems. It simply wouldn't run. After much tinkering, it later served to grind bark.

In 1795, a horizontal windmill (much like the ones built in the Middle and Far East), was running at Chipman's tannery in Beverly. Marblehead had windmills running at its saltworks as early as 1804. One was destroyed by a severe storm in October of that year. Another windmill, measuring one hundred feet in circumference and fifty feet high, was built by John Goodwin in 1819, near the old Marblehead Customhouse.

The flat, windy coasts and islands of southeastern New England were ideal for windmills. New England's notoriously blowy and changeable weather was harnessed less often inland, or where coastal rivers emptied into the sea. Thus, the greatest concentration of wooden windmills was along Cape Cod and the coast of Rhode Island, while comparatively few were used in Maine, Vermont, New Hampshire, and Connecticut. Still, tantalizing clues turn up on modern day maps and by roadsides—there is Windmill Point on Lake Champlain in Vermont; a Windmill Hill Lane in Castine, Maine; and a pair of ancient millstones set demurely in a suburban yard on Main Street in Old Saybrook, Connecticut.

Mark Twain knew and wrote about the vagaries of New England wind and weather better than any other writer. He saw how weather and terrain determined the region's economic well-being, but he found more humor in how the weather affected the taciturn New Englander's well-being. In his "Speech on Weather," given at the New England Society's Seventy-First Annual Dinner in New York City on Dec. 22, 1876, he said:

> *Yes, one of the brightest gems in the New England weather is the dazzling uncertainty of it. There is only one thing certain about it: you are certain there is going to be plenty of it—a perfect grand review; but you never can tell which end of the procession is going to move first. You fix up for the drought; you leave your umbrella in the house and sally out, and two to one you get drowned. You make up your mind that the earthquake is due; you stand from under, and take hold of something to steady yourself, and the first thing you know you get struck by lightning....*[4]

Twain wrote *A Connecticut Yankee in King Arthur's Court* and much of *The Adventures of Huckleberry Finn* while living in Connecticut. Windmills in and around Hartford, where he lived, are largely undocumented. Of nearby Wethersfield, arguably the oldest town in the state, The Wethersfield Historical Society states, "Lacking water power, windmills and dams were employed to process grain and cloth, and, in the Griswoldville section of town, to manufacture edged tools and run spindles."[5]

Though Cape Cod was to become the focal point for New England windmills, in the early years of settlement Cape Codders either ground their corn by hand or hauled it to Plimoth. Until the building of its first watermill in 1633, Plimoth itself relied on flour and meal imported from England at great expense, or on hand grinding. Stephen Deane was allowed in that year to erect a gristmill in Plimoth "for beat-

The Pligrims landing at Plimoth, 1620. Plimoth imported flour until a gristmill was built in 1633. Cape Cod relied on the Plimoth gristmill until windmills were built in mid-century. 19th century print by S.E. Brown.

ing corn." It was stated further that "the said Stephen can beat all the corn that is or shall be used in the colony and it shall not be lawful for any other to set up a work of that kind except it be for his own use." [6]

I remember one day in 1983 driving over the Sagamore Bridge—that high, arching entryway to the magical kingdom of Cape Cod. Just on the other side I saw about an acre of thatch being stitched to a roof, at one end of which was the skeleton of a new windmill. Was this a new museum of 17th century life on the Cape? After all, the Pilgrims had sheltered under thick layers of thatch in houses they had built in the 1620s, until they saw most of them burn down. Or was it an eccentric twist on the vast commercialization of the Upper Cape?

A reproduction windmill at the Christmas Tree Shop, Sagamore, Cape Cod.
Photo by the author.

I couldn't decide whether to be delighted or annoyed. This remarkable building became an enormously successful shopping outlet called, with no apparent logic, the Christmas Tree Shop. I was annoyed. Since then, though, I've had a change of heart. Perhaps the very first thing travelers to Cape Cod *should* see is a windmill, the unofficial symbol of the Cape. Some thirty-five to forty wind-powered gristmills are known to have been built on Cape Cod and the Islands. Hundreds more windmills were built to pump water for houses and farms, and to pump seawater for salt making. Only about a dozen windmills still stand on the Cape, many disguised as additions to houses. On the other hand, at least twenty-two lighthouses can still be seen on the Cape and Islands, giving an unfair advantage in the popular imagination to lighthouses.

The Christmas Tree Shop sells far more lighthouse-related souvenirs than it does those of windmills—even though the store itself is so closely associated with a mill. I tried in vain to find anyone in the building who knew much about the windmill. I bought a stained-glass windmill refrigerator magnet, and left. I have since found that it was built as a working mill and was meant to grind corn for sale at the shop. It is an example of just how complex and delicate the fashioning of a windmill is, for the millstones could never be properly balanced and it has never ground corn. It sits forlornly, but picturesquely, suffering the indignity of having its arms pushed around artificially by an electric motor. I have deep sympathy for this mill.

It was on April 1, 1644, that folks from Plimoth made their first permanent settlement on Cape Cod. Thomas Prence, a former governor of the colony, led seven families consisting of 49 people to the "Second Pilgrim Settlement," as town records called it. (Only Sandwich, Barnstable, and Yarmouth were settled earlier, and by people from other Massachusetts Bay towns.) This settlement was Nauset, now comprising the south end of Eastham on Town Cove. The total area they claimed also contained what are now the towns of Orleans, Eastham, and Wellfleet. Nauset retained its Native American name until 1651, when it was granted the name Eastham. Prence built the first meetinghouse here at the head of Town Cove in 1646.

During Nauset's early years, its corn had to be taken to Stephen Deane's water mill in Plimoth. After 1654, when Dexter's water mill was built, residents probably hauled corn to Sandwich, the town to the west, settled in 1637 by people from Lynn.

My question, of course, is this: When was the first windmill built on Cape Cod? Some of the Cape histories refer to a mill being built in the Sandwich-Barnstable area about 1633, but admit there is no documentation for this date. These same books report that Cape Codders before 1650 brought their corn to Plimoth for grinding. They certainly wouldn't have done so if a mill were available in the Sandwich or Barnstable areas. This supposed 1633 mill, later known as the Farris Mill, and no longer standing on the Cape, was likely built in the mid-1600s. Thus, it may or may not have been constructed earlier than the later windmills built in the 1850s in Eastham.

Eastham had no viable river or stream for waterpower, so a windmill was the logical choice. In fact, given this choice most Cape towns preferred windmills, for the damming of the "herring brooks" for waterwheels interfered with the spring run of alewives, the relatives of herring. Edwin Valentine Mitchell in *It's an Old Cape Cod Custom* (1949), told how this interference sometimes erupted in violent conflict: "In 1806 there was practically open warfare over this matter at Falmouth, where the free passage of the fish into Coonemosset Pond was prevented by a mill dam. The 'herring party' went so far as to acquire a cannon, but when it burst, killing the gunner, hostilities were discontinued."

The Harkness Windmill, with the miller and his family, Middletown, RI, 1900.
Photo courtesy of the Middletown Historical Society.

When a town did grant mill rights to a stream, these rights usually came with the condition that a run be constructed for the upstream migration of fish during the spawning season.

In 1650, Thomas Paine began to appear in Eastham town records. Paine was born in Kent, England, and as a young man had immigrated to New England with his parents. He appears again in the 1659 records, when he was granted four parcels of land. Thomas Paine was to become the most renowned of Cape Cod windmill wrights.

The first mention of a windmill appears in the Eastham town meeting accounts of 1660. Two men were chosen to mend "the common roadway lying at the head of the fields from the mill to the end of those bounds towards Harwich." In 1682, Eastham town meeting selected four men to negotiate with Paine for the construction of a mill at a part of town called Kaskaogansett. Paine was becoming much sought after. Two years later, in 1684, Eastham rewarded him with the following: "In consideration that Thomas Paine Sr. has been at grate charges about building two grist mills for the use of the town, the town hath granted unto the said Paine a parcel of 26 acres of upland with the township of Eastham." [7]

Thomas Paine, like a peripatetic windmill himself, moved all over the Cape to practice his craft. In 1661, Paine was granted permission to build a mill at Sautucket in Yarmouth. In 1687, he built Barnstable's first windmill, and was immediately requested to build another one for Yarmouth. In 1711, Truro asked him to build a mill near the Highland Light. Paine liked Truro so much that he never left.

This is a long way of saying no one knows when the first Cape Cod windmill was built, but it was most likely in the 1650s, most likely built in Eastham, and most likely built by Thomas Paine.

After Cape Cod, the highest concentration of windmills built in New England was located in Rhode Island. While the Providence area had good rivers for waterpower, Aquidneck Island and the Narragansett Bay area had more wind than fresh water. Windmills flourished from Narragansett to the west, through Jamestown and Newport, to Sakonnet to the east. By the late 1800s, there were some twenty windmills in Rhode Island, and the largest number were located in Portsmouth, where the first one was built in 1668 on Briggs Farm.

By the late 17th century, there were several windmills in Rhode Island, but apparently they weren't all of the best design. In the 19th century, William Emerson wrote, "Windmills may be seen on almost every eminence in this part of the country and in some instances they make a ragged and grotesque appearance. From their want of uniformity in their mechanism, it should seem that the best principles of construction of these machines so useful are not yet settled." [8]

Other contemporaries failed to appreciate the windmill's natural beauty—especially after the fearsome sails frightened their horses and sent them flying headlong onto the ground. The Rhode Island Historical Society Library in Providence contains the diaries of the Rev. John Pitman, Baptist Minister and ropemaker. On September 4, 1789, he described a carriage accident caused by a windmill: "My Horse ran away with the Shaize (carriage) and tore it to pieces. Went home on horseback."

Poor Rev. Pitman suffered two near-death experiences six years later. In 1795, a ship he was on collided with a sloop at three in the morning. Only three months later, he entered the following in is diary: "Sunday, Nov. 8, 1795. About 12 O'clock my Horse frightened by a Wind-Mill ran away with me, Mrs. Pitman, her niece Jacky, Betsy & Becky, in the Carriage, which damaged it but not us, going over a wood pile. I expected all was gone...."

It was, of course, ludicrous for farmers to build a mill, or buy one, for personal use; it was far more complicated than building a barn or plowing the earth. The art of the millwright was second only to that of the shipbuilder in difficulty, and there were few in the colonies who had mastered it. This adds one more reason for finding windmills near seafaring towns, for some shipwrights crossed over to making windmills. Similarly, millers themselves were at times former shipmasters, the skills for each being rather similar. In 1800, Freeman Atkins, for example, found the transition from shipmaster to miller fairly easy at the Truro windmill. The scarcity of good millwrights also helps account for the wandering of the mills. When a town needed one, it was often easier and less expensive to find one out of use elsewhere and move it, rather than locate and pay a millwright.

Mills were so vital to survival that they were exempt from taxes, and the miller was exempt from military service. Nor was the miller required to take a turn holding public office, and he was often given a choice piece of land adjacent to the mill. Millers earned a good living off "the miller's pottle," a stated quantity he was allowed of every bushel of corn he ground. This varied, but generally amounted to about two quarts from every bushel of grain.

It was far from an easy life. When corn was in season, the miller worked at full tilt. The wind being fickle, certainly less constant than the flow of water from a millpond, the miller sometimes worked through the night until the wind gave out. At the beginning of each "daily grind," the miller had to negotiate the long arms to set the sails, which were rigged nautical-style with halliards. When the wind blew hard, he had to shorten sail; and as the wind shifted, the mill had to be turned into the wind, like a ship tacking its way through the sea.

At the end of grinding, the miller, without the kind of crew found on ships, unfurled the sails, and put the mill to bed. Henry Hall, a retired sailor, had learned well at the mastheads of fishing schooners; but with no crew, he found himself in deep trouble one day at his Dennis windmill. As the wind picked up speed one day, he stopped the mill to shorten its sails. At the age of seventy, he was perhaps less attentive than during his younger days rigging sailing ships. Hall neglected to either set the brake or fasten the arms with the iron chains meant for the purpose.

Just when Hall had climbed out half-way onto one of the arms, a gust of wind caught the sails and the arms began to turn. Hall, in a deft move that kept him from being hurled to the ground, scrambled back along the revolving arm to the central horizontal shaft. He perched there, with the shaft turning under him, three stories above the ground. The exhausted old mariner was rescued by neighbors who spied him shifting his arms and legs, crablike, to keep the revolving shaft from dumping him into the spinning blades of the mill. ●

The Eastham Windmill on the Eastham Common, early 20th century. Originally built by Thomas Paine in Plimoth, 1680s.

Chapter Two

From Cornfield to Cornmeal, from Pure Wind to the Power Grid: *How Windmills Work*

"The most foreign and picturesque structures on the Cape, to an inlander, not excepting the salt-works, are the wind-mills—gray-looking, octagonal towers, with long timbers slanting to the ground in the rear, and there resting on a cart-wheel, by which their fans are turned round to face the wind.... A great circular rut was worn around the building by the wheel. The neighbors who assemble to turn the mill to the wind are likely to know which way it blows, without a weathercock."

Henry David Thoreau, *Cape Cod*

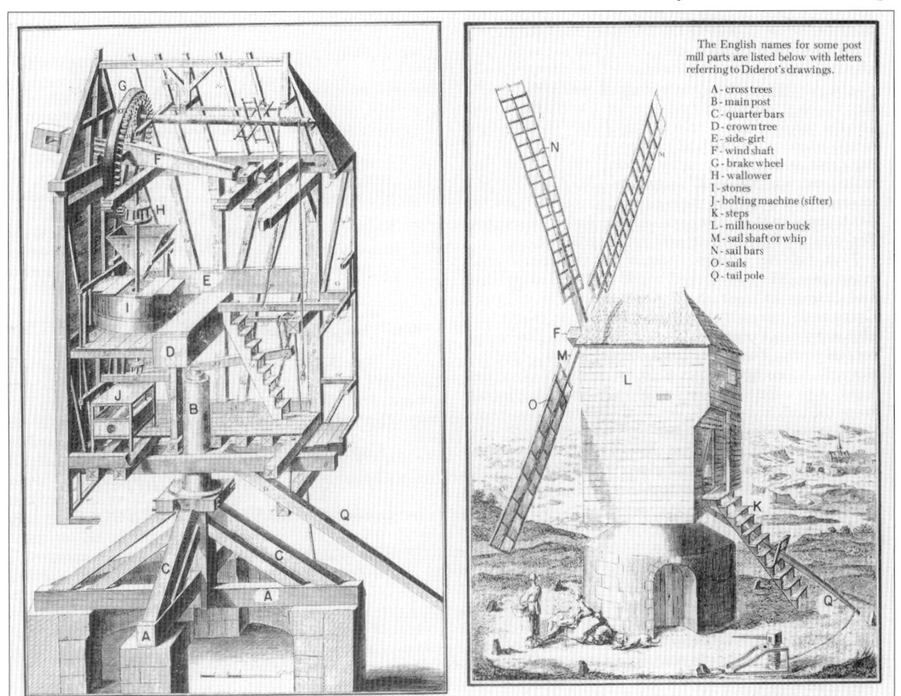

Engraving of the parts of a post windmill, Denis Diderot, 1766. Photo courtesy of William Marks, *Martha's Vineyard Magazine*, Summer, 1985.

The importance of bread to life is universal in the Western world. Bread is the staff of life. We pray for "our daily bread." We call our money, the very currency around which all commerce revolves, "bread." But it wasn't until the eccentric 19th century nutritionist Sylvester Graham came along that science took a cold, hard look at how bread was made. It was Graham—now known mostly for his cracker—who first claimed that how grain is milled and sifted is critical. What went on within the mill could lead to vibrant health, or a life that is "odious and abominable."

Graham lived in Northampton, Massachusetts, where he became the first American food reformer. His lectures sometimes caused women to faint (when he preached sexual theory), and bakers to riot (as happened in Boston when he told them their bread was no good). In 1837, he wrote:

> *Even the bread, which is the simplest form into which human ingenuity tortures the flour of wheat, is...too frequently rendered the instrument of disease and death, rather than the means of life and health, to those that eat it.... They who would have the very best bread should certainly wash their wheat, and cleanse it thoroughly from all impurities, before they take it to the mill; and when it is properly dried, it should be ground by sharp stones which will cut rather than mash it: and particular care should be taken that it is not ground too fine. Coarsely ground wheat meal, even when the bran is retained, makes decidedly sweeter and more wholesome bread than very finely ground meal.*[1]

The labors of the millwright and the miller were complex and difficult. Working primarily with wood and stone, they performed their tasks within very narrow tolerances. What happened, exactly, when a farming family brought their sacks of corn, wheat, barley, oats, or rye to the miller?

Before the windmill could begin to take grain into its hopper for grinding, the miller had to rig the arms of his mill with sails. With so many sailmakers along the coasts of New England, having suits of sails made for windmills was relatively easy, though expensive. Since many windmillers had been to sea, they found handling sails aloft a windmill relatively tame compared to clambering up the masts and rigging of a sailing ship in the gales of the Atlantic. As on any ship, the windmill's sails were trimmed according to the strength of the wind, and the windmill, in effect, was "sailed." On tower mills, the windmiller turned the sails of the mill toward the wind by turning the cap and arms, while the body of the mill remained stationary.

On most New England windmills, a tailpole extended from the top of the mill to a cart wheel which rested on the ground (some mill caps were turned by a chain wheel system). If the mill got caught in a tailwind, the sails could be blown out, the cap ripped off, or the entire windmill could capsize. Once the miller safely moved the tailpole, and the sails came into the wind, he was ready to release the brake.

The arms of the mill turned the horizontal windshaft within the cap. Mill caps were peaked like a Cape Cod roof, or boat-shaped (peaked with bulging sides like an inverted boat), domed, or conical. Within the cap, a brake wheel (or head wheel) was attached to the windshaft. This vertical wheel could be braked to stop the mill, but its main function was to turn the vertical shaft (or the "pinionshaft"), which was attached to a millstone below.

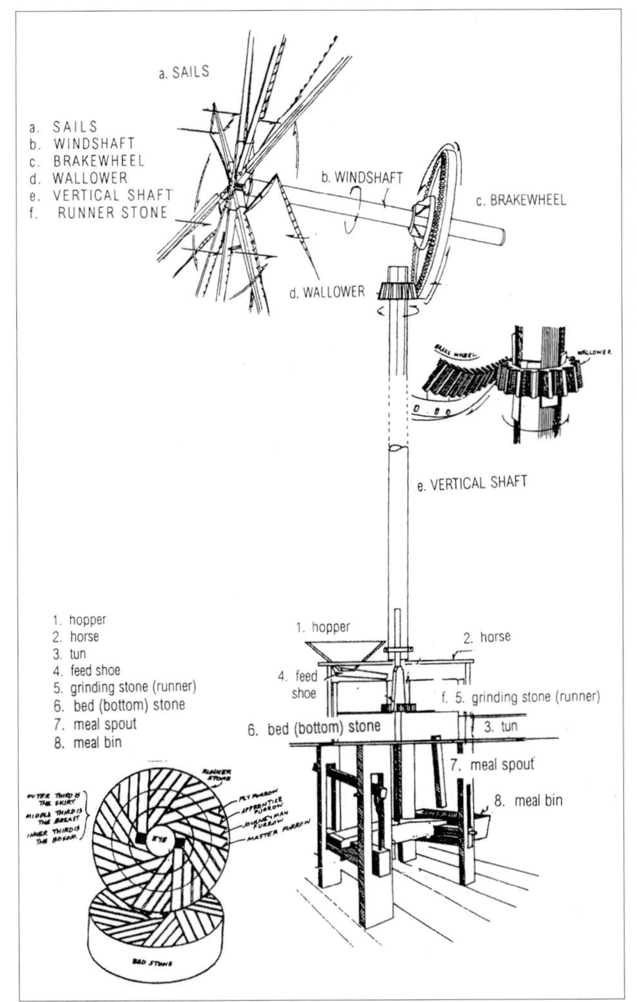

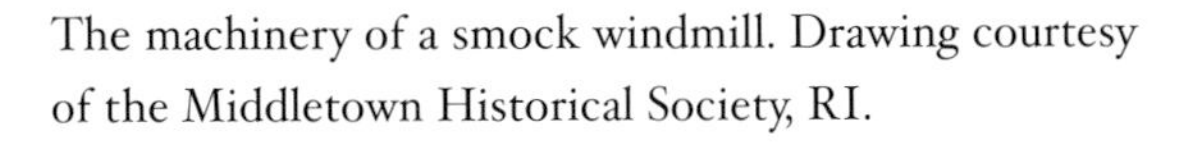
The machinery of a smock windmill. Drawing courtesy of the Middletown Historical Society, RI.

Woodcut of the hopper in the Old Mill, Nantucket Island. From *Harper's Weekly*, August 16, 1879.

The heart of the mill was the Stone Floor, usually the floor located below the level of the cap. Here, the grain was ground between the two millstones, together weighing 3,500 to 4,000 pounds. The top millstone, attached to the vertical shaft, was called the runnerstone, for it was the only stone that turned. The stone was raised or lowered to adjust the distance between the runnerstone and the stationary "bedstone" (or "netherstone") below it. The miller thus modulated the fineness of the flour and meal. Large mills requiring a more elaborate system of machinery might have positioned a Dust Floor above the Stone Floor.

The miller poured grain into a hopper and it trickled down a shoe (or chute) into the eye of the stone. The grain was swept to the outside edges as it was ground between the stones, which had been specially carved with one of several possible designs. Basically, the "furrows" were carved grooves which channeled the ground meal out of the stones and let in air to prevent fire. The "lands," which did the actual grinding, were the raised areas between furrows. The lands had smaller grooves called "stitching" carved into them. This shallow stitching wore away and had to be "dressed" or recut every month or so during active milling.

Woodcut showing the windshaft and brakewheel in the Old Mill, Nantucket Island. From *Harper's Weekly*, August 16, 1879.

The two stones were surrounded by a round wooden case, or "tun." Ground flour left the tun through the "spout" which directed it to the Meal Floor (or Spout Floor) below. On the Meal Floor, 800 to 1,000 pounds of flour were bagged on a good day. Beneath the Meal Floor, larger mills had another floor, used mainly for the miller's office and storage. Missing would have been a means of heat for, as mill historian Theodore R. Hazen explained, "A stove would not be exposed to the same space with the dangers of dust explosion. Flour dust is more explosive than gunpowder and 35 times more explosive than coal dust." [2]

Farming families would pick up their sacks at the windmill, leaving a certain amount for the miller as payment. This miller's "toll," or "pottle," was generally 1/16th of the grain a farmer brought in, or two quarts per bushel. This the miller sold to those who didn't grow their own grain: the blacksmith, shoemaker, ship's chandler, cooper, merchant, etc.

Woodcut of a post mill.

The evolution of the windmill illustrates a funny thing about the human personality: Sometimes it takes centuries for the obvious to take hold. Simply put, there were two basic types of European windmills: the post mill and the tower mill. Early Northern European millwrights built post mills, whose blades and machinery were contained in one box-like millhouse. The millhouse was held up by a single post, and the post was supported with trestles. This charming mill had two distinct problems: To get into the mill, the miller, often carrying heavy sacks of flour, had to climb a ladder or stairway. Moreover, when the wind shifted, the entire millhouse had to be turned. Within a few centuries, men tired of this.

Few post mills survive, for in the 16th century, the Flemish invented the tower mill. With a simple shift in thinking, it was decided that only the blades and cap of a windmill needed to turn, and all its gears, the tons of millstones, and other trappings could stay where they were.

There were two types of tower mills. The first were built of stone or brick and were circular in shape. Often a wooden gallery or stage was built partly up the tower to enable the miller to more easily get to the sails. In the 17th century, a new style of tower mill was developed, the smock mill. These wooden towers sloped elegantly along eight sides, lending them the appearance of a countryman's linen smock. With large oak trees for beams, pines for floors and sheathing, and cedars for shingles—all plentiful in New England—it was the smock mill design that proved to be the most popular.

Tower mill with gallery. From an 18th century millwright's trade card.

The various caps topping off windmills added to their distinctive regional styles. One of the oldest caps, the Kentish Cap, was common on the early post mills. It was a round vault, like a tube cut in half, with flat ends. This style was prevalent in Kent, England, and also in parts of Sussex and Essex.

The Sussex Cap was round and somewhat onion-shaped. It had a slight upward,

The Farris Windmill, West Yarmouth, MA, with a boat-shaped Norfolk Cap.

reverse curve that ended in a finial knob. Also known as "ogee shaped," it was also found in the East Midlands and northeast of England, and in the northeast of Denmark.

The Norfolk Cap was particularly aerodynamic, being shaped like the bottom of a boat turned upside down. Like a boat, the sides were broad and tapered to the ends. It was used in the Norfolk Broads of England, in Lancashire (in a larger version), and in the southwest of Denmark.

The Flemish Cap was cone-shaped with a dormer, out of which extended the windshaft and the arms. It's a style older than the Norfolk Cap, according to historian James Owens, and tops the windmill on the Eastham Common on Cape Cod. The Pent Roof Cap, also an early design, was a simple peak, like the roof of a house. Many Cape Cod windmills had the Pent Roof Cap, and some had either the Flemish or Norfolk styles. A plain Dome Cap, without the curves of a Sussex (or Ogee) Cap, was common on the windmills of Rhode Island.

When grinding was done, or when the mill was idle for any length of time, its sails were removed and the arms set diagonally. This resting position, called a St. Andrew's Cross, allowed the strain on the stocks to be equally shared. Because windmills were so clearly visible from great distances, they were sometimes used, like the sound of the church bell, to communicate with the local village. A particularly detailed set of signals were developed in Holland and other parts of Europe, some of which carried over to America.

A cone-shaped Flemish Cap on a Truro, MA, windmill. Photo courtesy of the Cape Cod National Seashore Archives.

A simple message, such as "the miller will be back soon," was con-

The Prescott Farm Windmill, Middletown, RI, exhibits the typical Dome Cap of Rhode Island windmills. Photo by the author.

veyed by setting the sails in a St. George's Cross, a vertical cross. If there was a death in the community, the sails would be stopped just prior to the vertical position. Mourning could also be illustrated by removing some of the boards on the arms—the more boards removed, the closer the relationship of the deceased to the miller. If the miller himself died, many crossboards were removed and the arms were turned slowly during the funeral.

Joyful occasions were marked by stopping the sails just after the upright position, and by decorating the arms. For weddings, brightly colored swags and garlands were strung between the sails. Patriotic celebrations were cause for hanging bunting and flags.

At night, lanterns were hung in windows or on the arms to signal ships at sea. An English legend tells of a windmill in Sussex that was used to signal smugglers when it

The Eastham Windmill, MA, windmill, with sails set in the St. George's Cross position.

was safe to come ashore with contraband silks, tobacco, and tea from France.

During wartime, windmills sometimes played important roles. When the British captured Nantucket during the Revolutionary War, islanders set the sails of their most prominent windmill to signal their ships not to come into the British trap. The British waited to capture these ships, but they never arrived, having been alerted to the danger.

Windmills were also used to signal airplanes in 20th century warfare. When the Germans occupied Holland during World War II, windmills sent messages to the Allied pilots. When they crashed in Holland, the Dutch underground helped them escape—communicating with each other through the language of windmills.

By the late 1990s, however, the goodwill and good language of the windmill, as exhibited by the Dutch during World War II, had dissipated. The "language of windmills" became more cantankerous than heroic. In New England, wind power was receiving a chilly reception in places like Block Island and Nantucket. The visual impact of the new windmills upset owners of seaside summer homes. Some objected to the sound of the rotor blades, others questioned whether birds, ocean life, and local aircraft would be sliced and diced or otherwise interfered with.

Meanwhile, Europe—and notably Denmark—began designing and installing windmills with an increased capacity to produce electricity. These wind turbines, with blades like aircraft propellers, turn electric generators which produce electric current.

Small models are used by individual home owners, farms, etc. Windmills are also grouped together into a single wind power plant. Wind farms, as they're known, are built both on land or offshore where wind currents are more reliable.

By the end of the 20th century, some European countries were producing as much as eighteen percent of their power needs from sleek new windmills. The United States made a half-hearted attempt to subsidize renewable energy, but continued to rely on cheap fossil fuel. One exception was a large wind farm which succeeded in California while oil prices were high, but was discontinued when prices came down. Undaunted, California again entered the wind power field in a major way. The U.S. Department of Energy reported that in 1990, California's wind power plants offset the emission of more than 2.5 billion pounds of carbon dioxide and 15 million pounds of other pollutants that would have otherwise been produced. It would take a forest of 90 million to 175 million trees to provide the same air quality.

Wind turbines at the Searsburg Wind Power Facility, VT, 2003. Photo by the author.

The largest modern windmills (one can be seen in Hawaii) have propellers that span more than the length of a football field, and stand twenty stories high. Each produces enough electricity to power 1,400 homes, according to the U.S. Department of Energy. Smaller, home-size windmills have blades between eight and twenty-five feet long, stand more than thirty feet tall, and can supply 8,000 to 12,000 watts of electricity, enough to power an all-electric home or small business. If these windmills produce more energy than is being used at a specific time, utility companies will purchase the extra electricity. This has been true since the passage of the Public Utilities Regulatory Policy Act of 1978, and works in one of two ways: Payment can be made according to readings from a separate electrical meter, or "net metering" can be used. With the latter, the meter turns backward while wind energy is supplied to the grid by the windmill.

Electric power windmills are, in some ways, simpler than traditional wooden windmills. They are comprised of rotors (usually called blades or arms on traditional windmills), an electrical generator, a speed control system, and a tower. Many are equipped with fail-safe shutdown systems, so that if the windmill fails or the wind

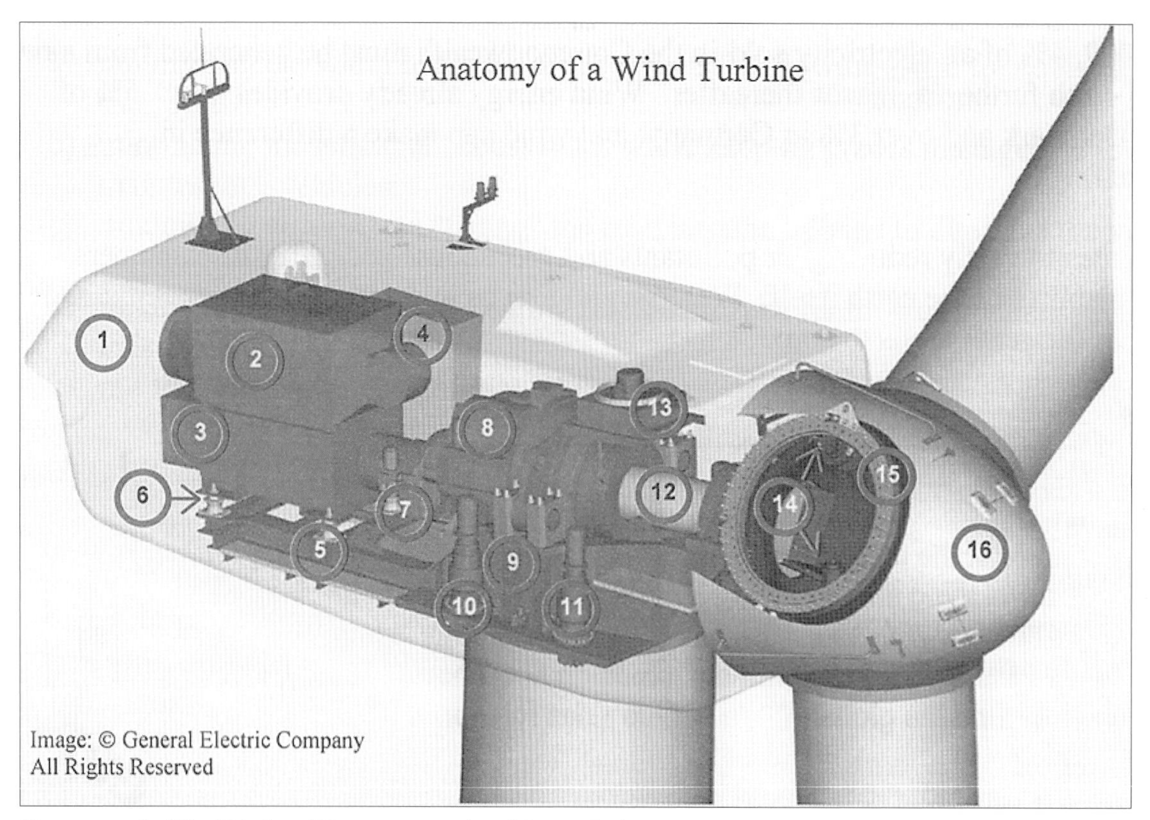

"Anatomy of a Wind Turbine." Image copyright of General Electric Company.

becomes too fierce, the system turns the blades out of the wind and applies brakes.

There are signs that well into the 21st century, renewable energy sources like the modern windmill may become as common as the old windmills of Holland. In January of 2003, the Associated Press reported that in Ede, Netherlands—a town with a classic wooden windmill—"a shiny new stainless steel windmill on the roof of a technical school barely whispers as its blades spin in a brisk winter breeze." [3] This is one of the new breed of urban windmills. Dutch cities including Amsterdam, the Hague, Tilburg, and Twente plan to install these urban turbines, each part of small-scale projects with fewer than a dozen individual windmills. The new windmills pay for themselves in about five years, according to the Dutch manufacturer, Prowin. Germany, Finland, and Denmark have also been developing windmills for urban centers.

Wind farms, like the Green Mountain Power Wind Facility in Searsburg, Vermont, are putting New England on the modern windmill map. Perhaps a 21st century Henry David Thoreau will one day look out over the New England landscape and be able to say, as Thoreau observed of the 19th century New England windmills, "They looked loose and slightly locomotive, like huge wounded birds, trailing a wing or a leg, and reminded one of pictures of the Netherlands." [4] ●

Chapter Three

The Surviving Windmills: *The Moon Was Up and That Air Was Sweet*

"This old mill has seen the little babe in its mother's arms, with its big eyes staring in wonder at the mill arms with their white sails swinging slowly through the air. It has seen this same tot grow into youth and manhood, has seen him become the head of a family and bring his grist of corn to mill, to be ground into sweet Johnny Cake meal to feed his little ones. It has seen his hair whiten, and his step falter, and finally seen him gather to his fathers. It has seen his sons bring their grist to grind and pass on, and still its big white sails swung slowly through the air as if defying time."

Benjamin F.C. Boyd, miller of Boyd's Mill, Portsmouth, Rhode Island[1]

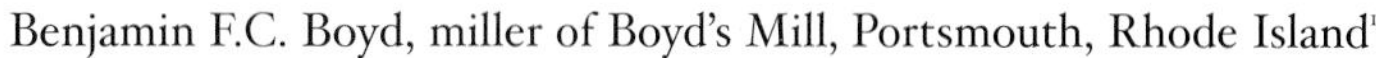

The Eastham Windmill, MA. Photo by the author.

The Eastham Windmill

Eastham Common, Route 6 and Samoset Road, Eastham, Cape Cod, Massachusetts

The Cape's most renowned windmill is the Old Mill on the Common in Eastham. It's the oldest operating windmill in Massachusetts (circa 1680s), and sits surrounded by just the right amount of daffodils, flowering shrubs, and elegant trees. The Lower Cape's only through-road, Route 6, funnels every visitor past this majestic mill.

For generations, the origins of the mill were lost. It is clearly very old, having a conical cap, in the older Flemish style, rather than the Norfolk boat-shaped caps of other mills. Like all of famous millwright Thomas Paine's known mills, it is of the smock type, an eight-sided tower tapering at the top. It was generally accepted that Paine had constructed it, and many logically assumed that he had built it in his home town of Eastham. However, recent thought, supported by miller James Owens and historian Fredrika Burrows, holds that Paine built the mill in Plimoth in the 1680s. Around 1770, it was dismantled for the first, but hardly the last, time and ferried across Massachusetts Bay to Truro on a log raft.

Meanwhile, Eastham was certainly not going without a mill. Windmills had been built there in the 1650s and 1680s. Another was built on the Common around 1705 by millwright Nicholas Paine (part of the Paine family of millwrights, and probably the son of Thomas) in partnership with Lieutenant Joshua Bangs. Called the Setucket mill, it stood for perhaps one-hundred years.

In the 1790s, when Eastham's Seth Knowles was looking for a mill to buy (rather than going to the expense of commissioning one), he found that the Truro mill was for sale. Knowles bought it in 1793 and moved it in pieces by oxcart. He reassembled it in Eastham on a hill next to Salt Pond, very near the present Visitors Center of the National Seashore. Later, in 1808, he moved the mill yet again: After selling the land to neighbors, he took the mill apart and moved it to its present location on the Eastham Common.

Thus, the mill built in Plimoth by Eastham's Thomas Paine completed a long loop of travel from the root of the Cape, out toward the end of the Cape, and back to Paine's hometown, in roughly 130 years. Just to tie things in an even neater package, in 1859, Joshua and Lucinda Knowles sold a 1/8 share of the mill to E. Thomas Paine, a descendant of the original millwright.

Jim Owens, the mill's present-day miller, likes to point out that in the 1800s the Old Mill was worth about $500.00, in an era when potatoes were 42 cents a bushel, gin was 39 cents for 1 1/4 pints, Hyson Tea was 30 cents for 1/4 pound, sperm oil was 20 cents a quart, molasses 34 cents a gallon, and beef 4 and 3/4 cents a pound.

In his book *Cape Cod Pilot*, Josef Berger described the mill in the days of Seth Knowles:

> *When she was in her prime she was a grand worker, and even now* [1937] *she can crush a lively bushel if the wind is right.... Kids watched her in awe as the sails filled to a brisk sou'wester and majestically cut great circles out of the sky; and they stood for hours waiting for Miller Seth to haul taut on his brakes and heave to; for then, if his "dispepsy" wasn't ailing him too bad, he'd let 'em go aloft, up in the cap, and see her great fan-shaft—bigger timber, it was, than a Grand Banker's mainboom—her giant hand-hewn gear and the massive spindle that turned the stone down on the grist deck.... To me, this old mill is an historical exhibit of the first rank—a bit of the dead past which lives and breathes still, which sings at her work and is close, endearingly close, to the earth and sky together.*

The Old Mill was a going business until 1896 when the Village Improvement Society bought it to convert into a library. This creative but molinologically damaging idea proved impractical. Instead, the mill was used to raise money to build a library farther down the road. At the mill, the Village Improvement Society sold homemade ice cream and littleneck clams (25 cents a bushel), and occasionally the Society ground corn. In 1928, the town bought and preserved the mill. It was first opened to the public in 1936, with John Fulcher as miller. Fulcher was the last miller to have started as an apprentice at the mill during its working days. Fulcher passed the craft on to others, like John Higgins in 1939, followed by Harold Cole in 1948, then Freeman Hatch, Jack Webster, and Clyde Eagles.

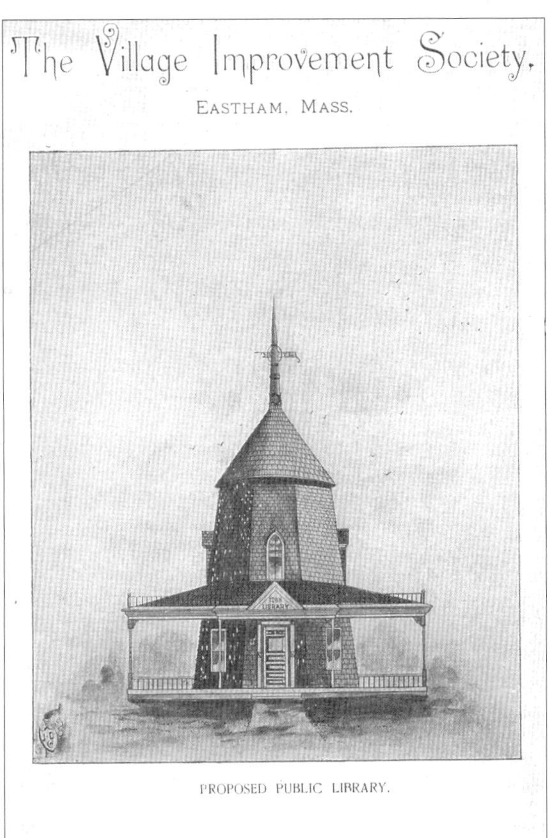

The proposed, but never realized, conversion of the Eastham Windmill into a library, 1896.

Windmill Green, as the common is now known, was redesigned in 1955 when the last remaining houses and buildings were removed from the site and landscaping was completed. Today, the Old Mill is in the expert care of Jim Owens, who keeps the mill open to visitors during the summer months and features it in the autumn during the town's annual Windmill Weekend. It's a tradition for many families to stop by at the mill every year to hear one of the Cape's great raconteurs. Here's Owens talking about windmills, as only one who has worked a mill can:

You couldn't ship flour from Boston down to the Cape, it took too long. It would go bad before it got here.... Stone ground whole wheat flour doesn't keep very well. The oils in wheat germ go rancid in no time at all, and with cornmeal it gets full of bugs unless you refrigerate it. They didn't have refrigeration so that meant you had to keep like a two week supply of flour around.... When you use it up you go back to the mill for some more. You bring the miller grain periodically, and he'd keep track of how much he owed you in flour, and how much his share was. Around here the rule was 1/16 of the grain was kept by the miller. And he could grind that and sell it. That was how he made his money. The shoemaker wouldn't have time to grow his own crop... and the tailor, the blacksmith, and so on.

A local miller was important, and the reason you found several in a town was because it took so long to go from one place to another.... So you have mills in South Wellfleet, mills in North Wellfleet, mills in the center of Wellfleet. There were four, five, or six mills in Orleans. Occasionally, you'd have one overlapping another, one went bust or got broken, and somebody built a new one. But all of the towns had them, and that's why they had them. They had tide mills and water mills and windmills.[2]

When Thomas Prence and the settlers from Plimoth established Eastham (Nauset), they brought their English farming practices and diet with them. They found the area to be one of the most fertile parts of the generally sandy Cape, although it was not quite as rich an area as Prence liked to claim: In 1634, he described the area as having the "richest soyle, for ye most part a blackish and deep mould, much like that wher groweth ye best Tobacco in Virginia."

Jim Owens talking to visitors at the Eastham Windmill, MA, Windmill Weekend, 2002. Photo by the author.

Eastham was sometimes called "the granary of Plimoth Colony," an appellation that, considering the shallow nature of the soil on Cape Cod, didn't carry much weight inland. Frederick Freeman, in his *History of Cape Cod* (1862), wrote, "...while other towns in the county can boast of superior farms, this town is the only one of the thirteen [on the Cape] that produces sufficient grain for home consumption." Within this sandy domain, Eastham did remarkably well, well enough in fact to keep the Old Mill on the Common active for many generations.

The Farris Windmill

The Henry Ford Museum, Dearborn, Michigan

The Old Mill in Eastham is the oldest standing on the Cape, but there is an earlier Cape Cod windmill in existence. It is located, to the dismay of many Cape Codders, in Michigan. Nearly 350 years old, the Farris Mill was wrenched away from the Cape in 1935, with its arms, figuratively speaking, flailing in protest.

The disappearance of the neglected Farris Windmill was a significant turning point in New England cultural history. Popular writer Joseph Lincoln was galvanized by the loss of the mill. Virtually singlehandedly, Lincoln created the 20th century image of romantic Old Cape Cod through dozens of books. Lincoln's writing inspired a small industry of windmill miniatures and reproductions on the Cape. Using the Farris Mill as the symbol

The Farris Windmill, West Yarmouth, Massachusetts

of the Cape's evaporating history, he called for the preservation of the fabric and image of the Cape. Thanks to Joseph Lincoln and the people he inspired, the loss of the Farris Mill kicked off the Cape's modern preservation movement.

Built in the mid-17th century, the Farris Windmill last stood on the Cape in a field at the corner of Berry Avenue and Route 28 in West Yarmouth. It is thought that some time in the 1750s, Lot Crowell moved the mill "from the north side of the Cape to the open field back of the old Ichabod Sherman homestead" along the Bass River in South Yarmouth. In 1782, David Kelley, Sr., and Captain Samuel Farris moved the mill from Bass River to a place called "Indian Town" (later called "Friends Village" for the Quakers who settled there). This move was described by Daniel Wing in his essay, "Old Cape Cod Windmills," in 1924:

> *The motive power, according to the testimony handed down from an eye witness, "was forty yoke of oxen," and help came from all directions, as may easily be imagined; especially as at a "house raising," there was likely to be plentiful supply of spirits for lubricating purposes. And there was; for the same eye witness further declared that "a barrel of rum stood on end with head knocked out; the contents free for all," with the natural result that, as the spirits of the workers rose, those in the barrel fell in like ratio, and "there was a noisy, drunken crowd in the vicinity that night composed of both Indians and white men."*[3]

In 1928, while describing the move, *Cape Cod Magazine* noted, "Painted in black on the inner wall of the old mill are the initials 'T.G.' and the date '1782.' The initials are said to be those of Thomas Greenough, an Indian...who took part in the moving and believed the occasion worthy of paint brush commemoration." [4]

There it stood for 112 years, in the care of Samuel Farris, his son Reuben, then his grandson Samuel; followed by Captain William Haffards; and finally Remegio Lewis, who had immigrated from Madeira, Portugal, in the early 1800s.

A fine description of the mill exists, as written by Amanda B. Harris in 1886 in an article called "A Windmill Pilgrimage":

> *It was "down on the Cape" that we saw one of the mills in operation, and were shown all about it. Such a rude, strong door with a wooden latch that must have been two feet long, such winding stairs, such heavy beams, such a tower and look-out, such a mealy, odd, picturesque, never-to-be-forgotten place! We were even given leave to bring away as relics for an antiquarian society, two or three of the crumbling, mossy shingles that had been sunning on its sides nearly a hundred years.*
>
> *The owner showed us how he managed, how he hooked the canvas sails to the great vanes, and told us that the long stick of timber outside, clamped with iron, and with a big wheel at the end was "the tail," and how they changed it about for the wind; and as we listened the vanes went round and round, and the corn in the hopper came out meal.* [5]

After Remegio Lewis died, his Sears heirs sold the mill to F.A. Abell in 1894. A letter in the archives of the Historical Society of Old Yarmouth, reads,

> *South Yarmouth Jan 2nd 1894*
> *Dear Cos. Elisa*
> *The mill is sold to be moved to West Yarmouth. Is there anything belonging to it that is not in it? I shall try to send or deposite for you $110.—please instruct me how to dispose of it*
> *Mary Nelson so called was buried yesterday*
> *I wish you and Mario yes and Rolie too*
> *A Happy New Year*
> *Yours*
> *S Sears*

As the mill was being moved to its final Cape location in West Yarmouth, an article appeared in the April 14, 1894 issue of the *Yarmouth Register*. Written by a Farris descendant, it's a fairly reliable account:

> *Its former history, previous to its being moved here, is veiled in obscurity.... Some think it was moved here from "Great Marshes"* [now West Barnstable], *others that it came from Falmouth, but nothing bearing upon the subject can be discovered in the county records, and the oldest people now living about here cannot give any information, other than it was here when they were children.*

The writer of the article was the great-great grandson of the mill's early owner, Samuel Farris. The article goes on to note some of the dangers of getting too close to the spinning beast:

> *In 1829, a horse owned by Reuben A. Farris was struck by the mill arms and disemboweled. The rent was sewed up and the animal was serviceable for several years afterwards.*
>
> *In 1856, a cow owned by Samuel A. Farris* [Reuben Farris's son] *was struck by the mill arms and died in consequence.*

By 1894, the writer continued, the mill was in poor shape:

> *The mill did duty satisfactorily up to within a few years, when there came a day when an attempt was made to start her agoing; and the wind was baffling and when it did fill the sails it was with fitful gusts and the grinding was irregular. The old mill struggled again and again, but its efforts proved futile. The miller, disappointed, sorrowfully closed the door, turned the old key in the ancient lock and left her alone, where for many a year the daisy and dandelion, together with the purple bloom of the potato have flourished around her base.*
>
> *As she now moves along with faltering step, like some condemned malefactor, dismantled and bound in chains, pulled along against her will, with her face turned wistfully towards the scenes of her youth, an air of dejection seems to spread over her features.*[6]

After Mr. Abell's death, the mill was sold to Dr. Edward F. Gleason of Boston and Hyannis. Then, in the early 1930s, automobile magnate Henry Ford passed his seventieth birthday and was fast approaching eighty—milestones with no obvious connection to windmills. But the automobile dealers of Cape Cod saw it differently: They wanted to present Ford with a birthday present that would be symbolic of Cape Cod, and landed on the idea of giving him the Cape's oldest windmill.

To Dr. Gleason's credit, he went to the townspeople and offered to sell the windmill and the land around it to Yarmouth as an historical site. To their eternal shortsighted embarrassment, the townspeople saw a moss-covered mill in need of repair—a relic that plundering vandals had recently visited—and decided to pass on the offer.

On November 9, 1935, it was announced in the press that the mill was sold and would be moved to Henry Ford's Wayside Inn property in Sudbury, Massachusetts. This set off a storm of protest, even before it was realized that the article was mistaken, and the mill would actually go much farther away, to the Henry Ford Museum in Dearborn, Michigan. There, it would keep company with a dovecote built in 1620, the oldest registered post office in America, and the shop where Henry Ford built his first auto in 1896. Scores of Cape Cod natives and summer residents declared that removing the mill would be an irreparable loss to the Cape. Wires from the West Yarmouth Village Improvement Society were sent to Ford in protest, but it was too late.

Ford's museum hoped to document the industrial evolution in America from agriculture to the automobile, and wanted to document the part the windmill played in this development. On November 15, a crew of workmen from the Ford plant in Somerville, Massachusetts, arrived with a truck, set up staging around the mill and began to dismantle it. More protests were sent to Ford as its arms were dismembered, and the ancient gears and beams were taken apart and loaded onto the truck. One letter, signed by fourteen members of the Old Colony School Superintendents Union, urged Ford to allow the mill to remain in its "native and natural setting." The *Boston Herald* lamented that, with the removal of the mill "which has attracted thousands of tourists every summer," rival Eastham would be able "to boast the oldest mill on the Cape." But the mill slipped through Yarmouth's fingers and was gone.

The Judah Baker Windmill

Windmill Park, River Street, South Yarmouth, Cape Cod, Massachusetts

In the years following the loss of the Farris Windmill, the price of secondhand windmills rose sharply. As for Yarmouth, the town redoubled its preservation efforts and today exhibits one of the Cape's finest and oldest remaining windmills. The eight-sided Judah Baker smock mill had been built in 1791 in South Dennis, overlooking Grand Cove. After moving at least four times, and undergoing three renovations in Yarmouth, it currently overlooks the water at the foot of Willow Street along the Bass River in South Yarmouth.

The mill was first moved by Captain Freeman Crowell II, who moved it from South Dennis to West Dennis, where it overlooked Kelley's Pond. On August 23rd,

19th century woodcut of Yarmouth, MA, with windmills.

1866, Braddock Matthews bought it and moved it to Bass River, South Yarmouth, not far from where it is presently located. Miller Allan Lewis ran the mill for Matthews and an investment group composed of South Yarmouth sea captains.

Like many coastal mills, the Judah Baker mill also served as a signal tower. When the miller spotted a packet ship arriving from Boston, he hoisted a flag on the mill, alerting people of the ship's arrival. Similarly, on the day before the packet was to sail, the miller would hang out a black ball to warn traders and travelers it was time to go to the boat landing.

In 1875, Seth Baker bought the mill and ran it until his death in 1891. This, as was true for many windmills in the 1890s when steam power was a more convenient source of energy, was the end of the mill's working years. Artists William and Alice Stone bought the mill early in 1893, and it became a favorite subject of their paintings. They suffered expensive repair costs when a storm sheared off part of the roof and the windshaft in

The Judah Baker Windmill, South Yarmouth, MA.

1916. The mill was moved again when Alice's brother-in-law, Charles Henry Davis, acquired it and moved it to his "House of the Seven Chimneys," also in Bass River.

The Judah Baker's last move was made in 1953, when the Davis family donated the old mill to the town of Yarmouth. The town moved it to its present-day location in Bass River Park, and made the first of many repairs to the building. Old-timers gathered and recalled hearing tales of the mill in the days when the sails still turned. One recalled that people inside the mill experienced the sensation of being "almost certain they were on a sailing vessel under a strong breeze, with all the tackle groaning and creaking."

In 1973, the town raised $32,000 and contracted with Raymond McKeon to completely rebuild the old mill. McKeon specialized in building houses using 18th century methods. As he kicked an oaken corner post on the mill, his boot went through the outer shell, and pulpy dust resembling brown sugar poured out. "Rotten clean through," he said. "They all are." He was able to salvage 60% of the original structure and reproduce the rest. The mill was saved for approximately another twenty-five years. By 1998, however, storms and vandalism had taken their toll. Yarmouth again came to the rescue, raising some $50,000, and winning a Massachusetts Historical Commission grant for another $50,000.

Architect Tobin Tracey and windmill restorer Andrew Shrake examined the damaged shingles and the fire damage caused by vandals. "You've got a 200-year-old machine here. It would be a shame to lose it," Shrake said. "What else are you going to leave for your kids, the town hall and a bunch of gas stations?"

The Old East Mill when it was located in Orleans, MA. Postcard courtesy of Mr. and Mrs. Stanley Snow.

The Old East Mill

Heritage Plantation of Sandwich, Inc., Grove and Pine Streets, Sandwich, Cape Cod, Massachusetts

Driving out of Sandwich center on Grove Street, I could see the fifteen-acre Shawme Pond on the left. On its near shore is the Old Town Cemetery, on the opposite shore the Thornton Burgess Museum, and at the end closest to Town Hall, the Dexter Grist Mill. In 1654, Thomas Dexter built the first known water-powered gristmill on the Cape. But I continued farther down Grove Street in search of one of Cape Cod's most beautifully situated windmills.

Grove Street winds under scrub oaks to Heritage Plantation, known for its vast gardens and Shaker barn filled with Americana. In 1968, the year before the opening of the plantation, the Old East Mill, the last remaining windmill (at that time) in Orleans was moved to the museum. Orleans, the hilly town so identified with windmills that one appears on its town seal, was temporarily left lacking a windmill.

In 1800, just three years after Orleans separated from Eastham, the front section of the Congregational Meeting House on Main Street was replaced. The discarded oak and pine that was still sound was used as framing and boarding for a new windmill on Snow's Hill, now the top of Great Oak Road. Among the five investors in the mill was Isaac Snow, the town hero twice captured during the Revolutionary War, who managed to escape and brought home the suggestion that the town be named after the French Duke of Orleans.

Snow and his partners, William Knowles, Samuel Smith, Edward Jarvais, and John Freeman, sold the mill to Deacon Abner Freeman in 1811 for $83.31. The windmill's first move came when Deacon Freeman sold it to his minister, the Reverend Daniel Johnson in 1819. Johnson relocated it to the hill above Meeting House Pond, which had its own landing from which the Long Island-Connecticut-New York packets carried products to market. This proved important during the Civil War when all the Orleans mills were busy grinding enormous corn yields and shipping the meal to the U.S. Army for emergency field rations for Union soldiers.

Isaac Sparrow bought the mill in 1828 and ran it during the Civil War. He ground wheat, corn, rye, and barley, and by attaching a salt grinder simultaneously ground the product of the Doane's-Roberts' Cove Salt Works. After the war, Sparrow sold the mill to master mariner Captain Joseph Taylor, who had retired from seafaring in 1866. After Captain Taylor's death in 1904, John M. Gundry bought and preserved the mill. His children delighted in camping in it until the 1938 hurricane blew off the newly restored arms and windshaft.

Postcard of **Old East Mill** at Heritage Plantation, Sandwich, MA.

In 1957, the mill was bought by Charles M. Campbell and moved again, this time to a spot on Barley Neck Road overlooking Pleasant Bay. Fortunately, the mill was in a fine state of preservation, with virtually all of its original machinery intact. Lester H. Bassett's restoration work was relatively easy, and shortly the mill opened for tourists. In the 1960s, in a scenario reminiscent of the story of West Yarmouth's Farris Windmill, Campbell offered the mill to the town of Orleans. It was quite an opportunity for the town; but the $40,000 price tag was more than the taxpayers could bear, and the town known for its windmills was soon left without one.

Unlike the Farris Windmill, the Old East Mill didn't leave the Cape. Heritage Plantation bought the windmill and hired a miller to run it. For years, Malcolm Johnson was the Chief Miller. He proudly demonstrated the grinding of corn and explained the workings of the mill during the summer tourist season. At present, the mill no longer grinds corn, and an electric motor keeps the old arms turning.

The Red Brook Mill, Cataumet, MA. Photo courtesy of John and June McCahill.

The Red Brook Mill

1094 Shore Road, Cataumet Village, Bourne, Cape Cod, Massachusetts

Bourne is a town of jarring extremes: Some of the Cape's worst traffic jams and commercial developments are here, yet it contains the spectacular eastern shore of Buzzards Bay and the graceful Cape Cod Canal. Within Bourne is one of the Cape's oldest sites, the Aptucxet Trading Post of 1627; but it is officially the youngest of Cape towns, having been incorporated in 1884. There are three remarkably authentic windmills (moved from elsewhere) located in Bourne, and one supposedly "authentic windmill imported from Holland." The latter is extremely handsome but is actually a mock-Dutch windmill. The story of this eccentric reproduction, with its ties to Grover Cleveland and Rip Van Winkle, will wait for a later chapter.

John and June McCahill welcomed me into their home and windmill in Cataumet, one of several villages within the town of Bourne. The mill is attached to their cozy house situated among the trees on Shore Road.

The McCahill's windmill, the largest ever seen on the Cape, measures fifty feet to the top of the cap. Known as the Red Brook Mill, it was built in a style unknown elsewhere on the Cape. Its large dome belies the fact that it was built in Rhode Island, where the Dome Cap—rather than the Norfolk Cap (boat-shaped) or Flemish Cap (conical top)—was common. Though the arms and grinding machinery are gone, John was able to show me the hand-hewn, mortised and pegged timber, and the original winding stair that leads from the 22-foot broad first floor up to the dome.

The Red Brook Mill was built in Bristol, Rhode Island, in 1797, and was moved to Fairhaven, Massachusetts, in 1821. Located just across from the whaling port of New Bedford, the mill was at that time used for sharpening harpoons. Zebedee Macomber owned the mill and willed it to his grandson, Perry Macomber, in 1850. Perry was a successful New Bedford merchant with big plans.

In 1852, Perry Macomber bought eighty-eight acres in Cataumet from Joshua Handy and developed the harbor and wharves there for packet service to New Bedford and elsewhere. He re-located his windmill and built water mills in the Red Brook section of Cataumet. Macomber hired master miller Nicholas Shick of Pennsylvania, and shipped in grain from all the ports of Buzzards Bay and the islands. From the bustling wharf, oysters, flour, white pine, and oak were shipped out.

Macomber himself operated out of New Bedford and hired people to look after his mills and wharves in Red Brook. It was in 1853 that he moved his windmill from Fairhaven to its new site. That winter turned out to be extremely severe, and Macomber found he could have the mill slid along frozen Buzzards Bay and across to Red Brook.

Not long after, the railroad arrived and the mill had to be moved out of the way to the north bank of Red Brook stream. In the 1860s, David Nye kept a diary in which he recorded the cords of wood he cut and loaded aboard the schooners at Red Brook. He also made entries each time he took corn to be ground at the mill there. Proud to say he voted for Abraham Lincoln, Nye didn't reveal much more. Most of his diary entries are as brief as this one: "December 21, 1865. Stormy. About home. Shelled corn. Went to mill." We know that the miller during this period was Captain William Ellis, for scrawled on one of the mill walls is the following: "Jan. 1, 1868, wind N.E. and strong. Grinder, Uncle Ellis, quite warm for this time of year."

The Red Brook Mill never made it to its 100th anniversary as a working mill, as some of the great ones did. In 1869, it was struck by a gale which twisted its arms and damaged the machinery. It seems the mill was never able to grind corn again. Macomber himself died in 1873. Historian Simeon Deyo reported the mill in ruins by 1890.

The mill moved once again when the Gammon family of Bridgewater bought it in 1905. The next year Ferdinand Gammon had cosmetic restoration done and moved it to his summer home on Shore Road, Cataumet, where it stands today.

John and June McCahill have been taking affectionate care of the Red Brook Windmill, as John says, "Since ten years before we landed on the moon."

The Rothery Windmills

9 Red Brook Pond Drive, 1090 County Road, Cataumet Village, Bourne, Cape Cod, Massachusetts

The first **Rothery Windmill**, 9 Red Brook Pond Drive, Cataumet, MA. Photo by the author.

Agnes Edwards Rothery can be thought of as the literary sister of Joseph Lincoln. Neither as prolific nor well-known as Lincoln, her books described the same romantic Cape Cod as Lincoln's, but Rothery allowed space for some of life's more bittersweet realities. Rothery is particularly interesting, because she lived in a windmill. In fact, her family owned two, both of which are still family-owned in Cataumet.

In 1900, successful Boston insurance man John J. E. Rothery bought an old Chatham windmill and moved it to his summer estate in Cataumet. Each summer the Rotherys, with five active, bright children, took the train from their home in Wellesley to Cataumet. The windmill, attached to their house, had three floors of living space for guests, and made a dandy playhouse for the children. Rothery very shortly had another windmill moved to the estate, and both of them can be seen in old photos, before trees and other houses obscured the view. The first mill is on Red Brook Pond Drive, and the second on Old County Road. Rothery's daughter Agnes used the estate as the setting for fictional works like *The House by the Windmill*, published in 1923:

> *For nearly two hundred years the mill had stood thus, and the shining bay and the single giant oak on the hillside that sloped to the pond were all that remained of the friends of its youth.... The venerable mill remembered when the meeting house across the road with its meek white spire pointed like a finger to heaven was built a hundred years ago. It remembered when the first grave in what was now the tranquil graveyard just beyond was dug more than a century back. And it remembered the many generations of millers that had lived and died in the story-and-a-half cottage—homelike still with its compact Cape Cod lines—which stood a few rods away.*

The first windmill that John Rothery moved to Cataumet in 1900 had been built in Chatham around 1730, near the present location of the Chatham Bars Inn. The mill is believed to have been used as a combination windmill and tanning mill before it was moved in 1830 to South Orleans, close to Arey's Pond. In 1870, Captain James H. Arey had it moved again, this time to Mill Hill behind the old Town Hall, now the Academy Playhouse. At that time, it was known as the Center Mill for its placement between the East Mill (now at Heritage Plantation in Sandwich) and the Jonathan Young Mill, which stood on land occupied today by the Governor Prence Motor Inn on Route 6A. To add a bit more confusion, the mill has also been known, not without logic, as Captain Arey's Mill.

When John Rothery bought the mill, he had it dismantled by Oliver and Lewis Eldredge of Chatham. The Eldredges moved it by rail and reassembled it in Cataumet. Total cost of the move was $322.52, including the Eldredges' daily wage of $2.50.

By 1926, the estate passed out of the Rothery family. From then until World War II, one could buy lunch, afternoon tea, or dinner in what had been the Rothery house and mill. It was called "The House of the Wind Mill Tea Room." Chicken and lobster dinners were a specialty.

By 1986, Ernest S. and Katherine C. Arsenault bought the house and mill (the rest of the estate, including the other windmill, had been sold off over the years). The Arsenaults transformed the first floor of the mill into a den for Ernest, the second floor into a studio for artist Katherine, and the top floor, complete with mill shaft, into a guest bedroom.

The second Rothery windmill is more of a mystery. Located on County Road across from the Methodist Church, it may have been moved there from West Harwich in the early 1900s. Many early photos show the mill standing by itself with its arms broken. Since 1991, the mill has been in the affectionate care of Carolyn "Callie" Connor and Walter "Bob" Connor.

Callie had grown up in New York but spent summers swimming and sailing on Buzzards Bay. "I lived for being on the Cape," says Callie. After marrying and raising two sons in Vermont, the couple didn't merely buy a house on the Cape, as so many do. The Connors bought a vintage windmill. No house, just the mill. As Bob put it, "I felt that as a summer place, we could survive in it for a while."

The Connors became fascinated by the story of the Rothery's mills after reading *The House by the Windmill*. Each year since buying the mill, Bob has tracked down and given Callie another of Agnes Rothery's Cape Cod books. Their dream of recreating a "house by the windmill" was realized in the autumn of 2001 when they completed a lovely two-story cape attached to the windmill.

The mill now serves as a charming guesthouse, when it's not being used by son Steve Connor as a studio for building handmade guitars. Steve, a highly accomplished craftsman, recently gave up his studio in Boston to move to the Cape. He was the perfect,

Guitar maker Steve Connor at the second **Rothery Windmill,** 1090 County Road, Cataumet, MA. Photo by the author.

enthusiastic guide for my examination of the old Rothery mill. An original millstone serves as a front step into the first, brick-covered floor. Walking in, I saw built-in bunk beds against the knotty pine walls. A very small kitchen and bathroom were wedged in, as well. A tight, winding stair took us to the second floor where Steve's computer keeps him in touch with suppliers of fine woods and customers who buy his guitars.

The third floor is truly magical. A double bed takes up much of the space, over which are dark, old, hand-hewn and pegged beams. Cubbyholes are arrayed neatly around the room, and over it all is the mellowed wood of the windmill's windshaft.

The Connors have, in a sense, retraced the steps of the Rothery family of one century ago. Here is Agnes Rothery describing a fictional version of her family as they moved into their windmill house in "Clovelly," her name for Cataumet:

> *When the 3:10 train from Boston stopped at Clovelly that June afternoon, nearly the entire population of the small hamlet was on hand to see the new arrivals.... The news that Ezry Swift had sold the windmill house—and the windmill, too, although what anyone wanted with that useless ruin was a mystery—to "city folks" had caused a tremendous excitement. The story-and-a-half cottage had been cleaned and put in order, the old furniture kept as it was, and the grass, or rather the weeds in the front yard, had been scythed. The windmill's broken arms had been mended and hung back in place, although they no longer moved, and glass was in the tiny windmill windows, and latches on the heavy handmade doors....*
>
> *There were no more millers now nor had there been for fifty years, but the windmill, with its quaint roof tucked in under the eaves...still braced its mammoth timbers with the mark of the axe upon them. The handmade pegs still held the eight gray sides together.*[7]

Agnes goes on to tell the story of the fictional Archibald Ryder family. As the five children grow over succeeding summers, the mill is always the backdrop to their enchanted lives. As they mature, their world becomes more complicated as their own hopes clash with the dreams and expectations of their overbearing mother. Some of the children find success, but the tragic early death of one son shakes the family out of its illusions. The book ends with the quiet death of the mother in the doorway of the old mill. The final, hopeful scene is of the grandson, Sam, sitting on his father's knee "in the open door of the windmill."

The John Smith Windmill

25 Harbor Road, Harwichport, Cape Cod, Massachusetts

If windmills left ghosts when they died, victims of gales and fires, Harwich would be haunted. There were so many windmills in Harwich, and their records are so sketchy, that many are just misty outlines and lost memories. The vast majority of Harwich windmills, along with the three water mills that had been located there, are gone. Fortunately, the oldest known Harwich windmill is still standing, lovingly cared for by the Blake family.

In 1968, F. Turner Blake, Jr., and his wife Mary bought the windmill on Harbor Road to the west of Wychmere Harbor. Mary Daniels Blake had summered in the area as a child. She kindly sent me the history of the 200-year-old windmill, along with a pair of vintage postcards of the mill:

> *The mill was built in 1792 by John Smith on the southwest side of South Street in Harwich Center. Julio Barrows bought the mill. Later it was moved to the other side of the street when it was sold to Lindsey Nickerson and his son James. When Lindsey Nickerson died in 1890 his heirs sold the mill to Dr. Alexis Julian, who sometime later had the structure moved to the east side of Wychmere Harbor.*
>
> *It sat there about sixteen years until he moved it to an undeveloped piece of land farther south, where it was used for a short while as housing for the wires that supplied electricity to the neighborhood.*
>
> *The Foster family from Cleveland bought the mill and surrounding homes. The estate was pictured in the July, 1925, "House Beautiful" magazine. The Fosters sold out to Herbert Clapp of Boston in the early 1940s, and the mill then moved again about fifty feet to the south, where the Clapps made it into a guest house about 1946.*
>
> *The next owner was O. H. Cilley of Philadelphia who added on more rooms. I believe this was around 1952. We bought the property from Mrs. Cilley in April, 1968.*
>
> *When it was first converted into living quarters, the first floor was made into a living room and the second and third floors into bedrooms. There is a small bathroom on the second floor. Apparently the bureaus were put in the rooms*

The John Smith Windmill, Harwich Port, MA. Photo by the author.

> *before the walls were installed by the narrow staircase, as we have been unable to remove them!*
>
> *We replaced the beds by having them built upstairs after the old ones were chopped to pieces. The walls present a challenge when hanging pictures as they slope so one has to put a nail on the bottom to hold them in place.*
>
> *Biddle Thompson of the "Snow Inn" Thompsons was the builder who transformed the empty mill into living quarters. The beams still have initials carved in them from the time it was empty. My sons have always been sorry that we didn't let them carve theirs!*
>
> *A full millstone is outside the front door. The other stone has been cut in half with one piece outside the kitchen door and the other piece makes the fireplace hearth.*

The *Harwich Independent* wrote of this mill in December, 1906, and referred to its earliest location as being "near the almshouse." The newspaper noted, at that early date, a wave of longing for early Cape Cod:

> *The old windmills are as picturesque relics as we have on the Cape. Their attractiveness appeals to the enterprising businessman who caters to the summer*

contingent. They are grabbed up as eagerly as old-fashioned furniture and they give an air of histrionic charm that appeals to visitors to these shores. The Cape Codder makes a good business stroke by this catering to sentiment, of which there is a preponderance among the tourists of this country. Anything that savors of the homespun days hits hard.

19th century photo of the **John Smith Windmill**, Harwich Port. Photo courtesy of Mary D. Blake.

In 1937, writer Josef Berger praised the Blakes' mill in his *Cape Cod Pilot* as one of the last remaining and best kept of Cape Cod windmills. He lamented, however, that some "wealthy summer residents" were buying old mills merely as adornments and filling them with "lawnmowers, bridge tables, children and other impedimenta." Berger noted that, on the other hand, here was a splendid mill that had once "weathered neglected on the town poor farm," and would "doubtless command several thousand dollars now, if she could be bought at all." Indeed, this mill is a jewel that could fetch more than the price of all the almshouses ever built on the Cape.

Old Higgins Farm Windmill

Drummer Boy Park, Main Street, Route 6A, West Brewster, Cape Cod, Massachusetts

When Brewster split off from Harwich in 1803, the new town had a lot going for it. By 1849, John Hayward's *Gazetteer of Massachusetts* was able to boast, "From three ponds in this town, covering about one thousand acres, a never-failing stream is produced on which are a paper-mill, carding, and grist-mill." Brewster had a thriving salt making industry (thanks to the windmills along its shores), a mackerel fishery, and since 1795, a fine windmill to grind corn. According to some critics, Brewster had one more thing, however. A bad attitude.

Historian Henry Kittredge wrote in his *Cape Cod and Its People* (1930), "An apparently hostile attitude toward strangers took deep root in the Cape soil. In fact

Old Higgins Farm Windmill, West Brewster, MA. Photo by the author.

it never entirely disappeared. As late as 1810, the selectmen and town clerk of Brewster remonstrated against the appointment of Edward O'Brien as postmaster, 'he being a foreigner a catholik and, in the opinion of the town, an alien.'"

Frederick Freeman saw it differently. In his classic *The History of Cape Cod* (1862), he said, "Brewster is, in fact, one of the most agreeable towns on the Cape. The neat and thrifty aspect of the dwellings has long been proverbial." Freeman kept his gaze from wandering into the windows of the town's neat houses and churches, and thus avoided any hint of strife and controversy. Brewster outgrew its early reputation for surliness, and welcomed the 19th century migration of Portuguese to Cape Cod.

Many of the townspeople of Brewster had already traveled to the far ends of the earth and mingled with other cultures. Captain William Freeman was familiar with the South Seas and was an early visitor to Pitcairn Island, refuge of the *H.M.S. Bounty* mutineers. In 1883, he brought his daughter, Clara, to Pitcairn, where she formed a lasting friendship with island-born Rosa Young. For Rosa, granddaughter of mutineer Edward Young, Clara was the first white woman she had ever seen.

Captain Josiah Knowles also saw Pitcairn Island, but only after surviving a shipwreck. His ship, *Wild Wave*, wrecked on the island of Oeno, about 100 miles from Pitcairn. He set off in a small, frail boat, carefully guarding $18,000 in gold. Approaching Pitcairn, his boat was stoved in on the rocks. He built another boat, made sails from a shirt and pair of overalls, and proceeded to sail an impressive 1,500 miles until he was picked up by an American sloop of war.

One hundred years earlier, around 1795, the Old Higgins Farm Windmill was built in Brewster at the head of Ellis Landing Road. Little is known of the origins of this gray-shingled mill with its elegant, boat-shaped cap. However, we do know why it had to be moved from its original location: The creaking and cracking of its gears and the whirling of its sails scared the heck out of passing horses. After a number of bolting horses and broken wagons, the mill was moved from the busy road (now Route 6A) closer to Ellis Landing.

19th century woodcut of the town of Brewster.

Like most mills, this Brewster mill was owned over time by small corporations. In 1845, for example, it was owned by the Higgins family and several members of the Crosby family. In 1890, it was moved across Ellis Landing Road to the Roland Crosby Nickerson estate. This made it eligible for a hypothetical list of "windmills put to unusual uses," for the Nickersons used it as a clubhouse at the 9th hole of their private golf course.

In 1973, Frances Nickerson donated the windmill to the grateful Brewster Historical Society in honor of her late husband, Samuel. Her request that it be moved to a suitable location and be open to the public led to a mobbed emergency meeting of the Historical Society at Town Hall. The first line of an article in the *Cape Codder* said forlornly, "...Brewster people are looking for a place to put an armless old windmill."

The Drummer Boy Museum is just down the road from Ellis Landing on Route 6A in West Brewster. The museum's owners, Mr. and Mrs. Lewis A. McGowen, gave the Society more than an acre of lovely meadow, and in 1974 the old mill was moved.

The town hoped for a parade with brass band and floats, but the police chief and the selectmen pointed out a few slight problems. Carting the windmill would involve elevating utility lines, trimming tree limbs, and halting busy 6A traffic. A stop-and-go parade surrounded by frustrated motorists wasn't what anyone had in mind. Instead, the cap of the mill was moved to the new site, followed the next day by its body. The road was lined with Historical Society members who each held a stack of cedar shingles. For twenty-five cents, one could sign a shingle that would go on the mill, and the money raised would help fund its restoration.

Raymond McKeon of Yarmouth Port, the man who had won praise for his work on the Bass River mill in South Yarmouth, restored the Old Higgins Farm Windmill. McKeon found unusual ship's carvings on the feed bin reminiscent of sailors' ivory scrimshaw. Next to the stairway, an author, known only by his or her initials, had written the following sentiment:

Old Mill
your dear arms seem to
stretch out in loving embrace to me.
T.P.F. Dec. 26, 1893

Sur Mer Windmill, Chatham, MA. Photo by the author.

The Sur Mer Windmill

157 Shore Road, Chatham, Cape Cod Massachusetts

The most dangerous windmill on Cape Cod has to be Barbara Townson Weller's Sur Mer Windmill in Chatham. With some mills, it's the whirling arms and tons of spinning millstones that are hazardous. For Sur Mer (French for "on the sea"), it is the traffic-stopping, heart-stopping setting that cries out for crowd control and a defibulator. From busy Shore Road, a tranquil wildflower meadow with coreopsis and daisies slopes down to a small windmill set in thick grass. Just beyond is turquoise Chatham Harbor, with a perfect red fishing boat anchored nearby. To the left, a weathered saltbox completes this irresistible scene.

The main house of Sur Mer is known as the Joshua Bearse House. According to the Chatham Historical Society, it was built in 1820 on the corner of Silver Leaf and Water Streets in Old Chatham Village. The Seth Bearse windmill by the water was erected around 1850 on the north side of Cockle Cove Road in South Chatham. Windmill collector John E. Rothery of Cataumet bought the mill, but unlike the mill his daughter Agnes wrote about in *The House by the Windmill*, he didn't move it to his property. After Rothery's death, the Morse family bought the mill at auction and moved it to Sur Mer.

The windmill was re-set along the property line next door to the home of popular Cape Cod writer Joseph Lincoln. It appears in the background of countless postcards of Lincoln's place and is usually assumed to have belonged to him. It has, however, been lavished with affection by the Townson family since the Depression.

Barbara Townson Weller and her daughter, Leslie Weller, graciously let me briefly into their lives to get to know life with the windmill at Sur Mer. On a sparkling May morning, I walked past the little windmill to meet them in the old Seth Bearse house by the water. We sat around the dining room table. Before us were old photographs, scrapbooks, and numerous magazine and newspaper articles about Sur Mer. At one end of the beamed and whitewashed room was a cabinet with a collection of souvenir postcards, calendars, etc., all with the image of the Sur Mer windmill. Beyond the wide bench-seat picture window, we could see windmill and the harbor.

I asked Barbara to tell me the story of Sur Mer:

> *To the best of my knowledge, the windmill, this house, and a second house were moved onto the property in 1911 by two women. One was an American woman by the name of Marian Morse, and the other was a French gal by the name of Rachel Riordon. They made a tea room of the upper Brooks House, where the flower meadow is now, and they made a little tiny antique shop of the windmill. They built the addition onto the windmill to make it a kind of guesthouse/antique shop. Then they, as Chatham became a resort, rented out the upper house and that is how my family got here, on a fluke.*
>
> *Then my family made the decision to buy the place. My father bought it—I think it was $35,000, it may have been $30,000—two weeks before the Crash in 1929. Now mind you—there was no electricity, no plumbing. We had an outhouse, sort of connected to the Brooks House. I can remember every bathroom that has gone into the place. An honest-to-God working windmill was over the well to pump water.*
>
> *I suppose the most famous vignette about the mill—at least within our family's history—is about Thornton Wilder. You see, I had an extraordinarily difficult uncle, my father's younger brother, whose name was Mutt Townson. His real name was Andrew J. Townson, Jr. My grandmother Mary Antoinette had been widowed for so long, that when it came time to educate Mutt, she simply threw up her hands and said to my father, "Douglas, you'll have to be in charge of it."*
>
> *Mutt was brilliant, never cracked a book, and bad news! Very colorful, but bad news. He was kicked out of one boarding school after another on the East*

Coast, and finally my father, in despair, thought the way to do this was to get him tutored all summer, and then we can get him into a decent school and maybe he can keep up, if he elects to.

So Pa went to Boston to an agency that all the colleges used to place their students in jobs for the summer. A guy from Princeton by the name of Thornton Wilder applied for the job of Mutt's tutor. And Thornton lived in the windmill for at least two summers. My mother would look you dead in the eye and say, "Of course, Thornton wrote 'The Bridge of San Luis Rey' in the mill." He didn't, but he wrote the outline while he was trouping around after Mutt Townson.

It was absolutely outrageous. The second summer of tutoring Mutt, father or Thornton figured out that Thornton would take Mutt to Europe for the summer. So over they went and I think the minute they landed Mutt disappeared. Thornton had it worked out that they were going to go to cathedrals, that sort of thing.

As a child, I must have been a shadow to Thornton Wilder, because he gave me a goat and a goat cart. And I could lead the goat and cart around Sur Mer. I thought it was perfectly wonderful, and I had one doll, and I've never liked dolls. I put the doll in the cart that the goat was pulling, and then the goat ate my doll, and I have no idea what happened to the goat! My mother didn't like animals much and I was crazy about them. I think she gave it to the milkman or somebody.

Sur Mer Windmill, Chatham, MA. Photo by the author.

When we were growing up, the writer Joseph Lincoln had the house at the end of the driveway, next door. We used to call him, as his granddaughters did, Daddy Joe. He was most delightful...he was kind of like Santa Claus in khakis.

Many of the postcards that you see were taken from Daddy Joe's garden. There would be our windmill in the background and they'd say, "View from Joseph Lincoln's Garden." One day, my brother and I couldn't figure out anything else to do, so we set up a table on Shore Road, and we made a sign saying, "Tours of Joseph C. Lincoln's Windmill. 5 cts." And that lasted about two hours, and then we got "hooked" by the family. But we thought it was wonderful and had quite a turnout.

The Sur Mer windmill barely escaped the fate of the Farris mill from Yarmouth, which ended up very far off-Cape, in Henry Ford's Greenfield Village museum in Michigan:

> *This period would have been, I am betting, in the late '30s, before the war. A representative of old Henry Ford's came down one day and said they'd like to buy our windmill to take out to Greenfield Village. They offered father $10,000, and he refused. He always said afterwards, with his Scottish soul, "That was probably the dumbest thing I ever did."*
>
> *I rent out the entire place, and it works, though I don't tell renters that the mill is either hot and damp or cold and damp! Either they walk into it and say, "This is the most charming thing I've ever seen," or they say to themselves, "Let me have a jacuzzi at the Holiday Inn." Living in the mill is like living in a boat. There you are, there's the water, there's all the sound of living on a boat. But you've got to like it because it makes a lot of noise and skunks go under it.*
>
> *During Prohibition, rum-runners landed their booze here and walked up between the windmill and the houses. My mother described looking out her bedroom windows and saying, "My God, it looked like Treasure Island!" All the cases of booze coming up.*
>
> *After I was married, we used to freeload here when my father was alive, because we couldn't possibly afford to rent anything. We would have my daughter Lili, her brother Tim, and a keeper of some description stay in the Brooks House. My husband Gordon and I would be in the windmill, and the rest of the family would be here in the old house. Just about a perfect setup. As Gordon pointed out, the only thing we didn't have in the windmill was a refrigerator. All our friends would come back with us at night and we'd go to the windmill. Tiptoe over here to the house, get the bucket full of ice, bring out a bottle of scotch and sit over there and let her rip.*
>
> *We always had my daughter Lili's birthdays in front of the windmill, around the millstone. We had little chairs, newspaper party hats. Maggie the cook would come up with a cake with a marshmallow frosting and multi-colored jimmies. Lili hated it. She preferred wild blueberry pie.*
>
> *You know, Dan, the North Beach was not at all touristy then. The boat was in front of the windmill. We had this marsh. We picnicked almost every day. We'd just put the kids in the boat and go over to the North Beach. It was a wonderful existence. The last day of the season was so depressing. I'd walk down here and the children would be standing on the wall, just looking out into the water, and I'd think, "Oh my God, here comes the leaving blues."*
>
> *Now our grandchildren are growing up here—Tim has three children. Of the three, this firstborn daughter, Katie, she's got that same sand-in-the-shoes*

kind of thing. I'll see her walking by herself on the beach, picking up shells. Our kids are still waterbabies.

Mrs. Weller and I walked across the lawn to the windmill. She led me through the door into the first-floor room with pegged beams. Against one raw, wooden wall was a fancifully painted wooden toy chest, decorated by the well-known folk artist Peter Hunt. Hunt was a friend of F. Scott and Zelda Fitzgerald, and a friend of Mrs. Weller's mother. He lived in the windmill while decorating a bedroom on the second floor of the Bearse House.

Barbara Townson Weller in the doorway of the **Sur Mer Windmill**. Photo by the author.

This is an extraordinarily corny thing to say, but I believe it with all my heart: There are vibes in both this windmill and the Bearse House. This is a very happy place. A lot of good things have happened here. A lot. I have no idea why it is. I think because this house and mill have been loved and they know it. This place has put people through hell, me included. I always walk down at first and I think, "Ha! I've gotcha! We've got a new septic system, we've gotten the shingling done. We're up to scratch." And it's almost as if Sur Mer is thumbing its nose at me, and here we go again. It's an on-going love affair. And you'd have to love it or you wouldn't put the money into it.

The Halloween Storm of the 1980s washed away a bunch of camps across the water, at North Beach. Down in our marsh we had stoves and propane tanks... everything came across the harbor. It cost thousands to clean up the debris. The hurricanes since I've been here.... One took off the arms of the windmill.

Any problems that I have had in my life...if I can get down here...they take care of themselves. There's a healing quality if you love it and it loves you back. I was walking along the revetment last night about quarter to nine, and that moon was up and that air was sweet. And I thought, I am so lucky...you know, to have this. A friend of mind said, "You know, I think Sur Mer is a toy for you." And I said, "No, it's not, it's a way of life."

My feeling for this place...I wonder if you should love something that isn't alive as much as I love Sur Mer. Is there something sort of weird about that? Lili adores it. But Tim has that extra dimension about it, that my father had, that I have, that Katie has. I know the answer will come to me as to what to do with this place. I have said to both kids, "Look, when I die, Sur Mer has been it for me. Don't feel obligated, don't let it own you, because it will." But God, I've enjoyed it, right to the hilt!

Benjamin Godfrey Windmill, with blades in foreground, Chatham, MA. Photo by the author.

The Benjamin Godfrey Windmill

Chase Park, off of Cross Street, Chatham, Cape Cod, Massachusetts

Chatham has seen at least eleven working windmills since the 1700s, with as many as seven grinding at one time. Today, Lou Springmeier is the "Mill Keep" of the Benjamin Godfrey Windmill in Chase Park, between Oyster Pond and Mill Pond. One of the few mills left that has been renovated to full grinding condition, it has suffered the same fate as most others—the lack of skilled artisans left to run it. Springmeier does an outstanding job interpreting the mill, having been trained by Jim Owens. "It's like being in the old *USS Constitution*, when I see the size of these beams," he relates. It would take at least two millers to run the mill today, one outside and one inside. "Can you imagine a kid getting his head knocked by one of those sweep arms? His head would end up in Oyster Pond!"

Historian Daniel Wing noted in 1924 that when Benjamin Godfrey built his mill, John Adams had just become the second president of the United States. In 1797, it was erected on what became known as Mill Hill, overlooking Stage Harbor. Godfrey had fought in the American Revolution and had helped repel a 1782 attack on Stage Harbor by a British privateer.

Godfrey's nephew, Christopher "Uncle Tap" Taylor, inherited the mill and ran it from Godfrey's death in 1819 to about the close of the Civil War. As in the history of many Cape Cod towns, Eldredges and Nickersons then enter the story. Oliver Eldredge bought the property after Taylor's death, and Zenas Nickerson's family bought it when Eldredge died in 1874. Zenas was its miller for nineteen years and also ran a grain and grocery store. His son, George, followed him and ran the mill for four more years.

Charles Hardy owned it at the turn of the 19th century, an era that marked the beginning of a long stretch of assaults by Mother Nature, who seemed determined to destroy the old mill. In 1907, a northwest gale blew off and completely demolished all of the arms and the outer end of the mill's shaft. On April 8, 1924, the *Chatham Monitor* reported, "The gale of last Tuesday night and Wednesday did serious damage to one of our most favored landmarks, the old mill. It struck it in the tail and so tipped the head out, wrecking the trundle top and one arm. Mr. Hardy has had it inspected and hopes to be able to repair it."

Later articles in the *Chatham Monitor* note that a lightning strike left considerable damage in 1927, and in January of 1932, a "60-mile gale which was sweeping the coast hereabouts blew down the mill's large fan."

In 1939, the windmill was acquired by Stuart M. Crocker as an addition to his summer place on Mill Pond, Stage Harbor Road. Like other Cape Cod towns during this period, Chatham had an active Historical Society and a deep interest in preserving its material past. Crocker worked with the Historical Society on a plan to offer the town the Godfrey Windmill. In 1956, Chatham accepted his offer and, at Crocker's request, moved the mill to its present location in Chase Park, off Cross Street.

Historian Daniel Wing wrote about the mill in his 1924 pamphlet, *Old Cape Cod Windmills*. He recalled, "In our youth we greatly enjoyed the humorous story of Don Quixote's planned attack upon a group of innocent windmills which he mistook for giants; but really, a windmill when in vigorous action, is no mean antagonist, if one ventures too near its huge, swinging arms." Wing goes on to mention "victims" of Cape Cod mills, including a slain horse and cow, a man maimed for life, "and at least one boy wounded while engaged in the favorite pastime of running between the revolving arms."

Wing ends his little piece on the Godfrey Windmill with this tale:

> *While writing this series of articles on old windmills, the ancient story has frequently come to mind of a certain legislator who had gained notoriety for his dry, long-winded speeches to the extent that he had become a bore to his hearers.*

On one occasion, the length and dryness of his harangue had so parched his tongue that he ordered a page to bring a glass of water. Just as he raised it to his lips, a fellow member arose and in a very dignified manner addressed the chair:

"Mr. Speaker, I rise to a point of order."

The chair: "The gentleman will state his point of order."

The member: "Mr. Speaker, my point of order is this: that it is highly improper to run a windmill by water power."

Jonathan Young Windmill, Orleans, MA. Photo courtesy of Mr. and Mrs. Stanley Snow.

Jonathan Young Windmill

Town Cove Park, Route 6A, Orleans, Cape Cod, Massachusetts

When I went in search of the wandering past of the Jonathan Young Windmill, I expected to learn about sea captains, but not about bright red Lotus sports cars.

The history of the Jonathan Young mill starts out in the normal way. It was one of at least nine mills in Orleans which ground corn and grain from the farms, and rock salt from the local saltworks. Built some time in the mid-1700s, it first appeared on a 1798 survey map of Kenrick's Hill, now part of the watershed in South Orleans.

Jonathan Young was a successful Orleans businessman who had built a boot and shoe store into a large variety store. He also served as clerk and treasurer of the Cape Cod Central Railroad. The richest man in Orleans, Young decided in 1839 to add a windmill to his business interests. Along with William Mayo, Joseph K. Gould, Francis Young, and David L. Young, he bought the South Orleans mill. It was moved to a hill on Jonathan Young's property near Town Cove, the present site of the Governor Prence Motel. After a number of years, Young decided to sell the mill.

Meanwhile in Hyannis Port, Captain Henry W. Hunt was winning fame and fortune through his whaling fleet, which sailed out of the West Indies. Upon his retirement, Hunt took up farming in Hyannis Port at the corner of Scudder Avenue and Pitcher's Way on property he had purchased from Lemuel B. Simmons, a prosperous sea captain. Rather than build a mill, Hunt purchased Jonathan Young's old mill in Orleans. He had it dismantled and moved by oxen to the shore of Town Cove, then by barge over the waters of Nantucket Sound, and then, again by oxen, to his farm in Hyannis Port. It was during this period prior to World War I that the working windmill began to appear in color postcards.

The windmill passed through the hands of Manuel H. Lombard of Winchester after Hunt's death in 1919. Lombard's daughter, Mrs. Robert Groves, inherited the mill around 1978, and offered it to the Orleans Historical Society in 1983. The Historical Society arranged to donate the mill to the town of Orleans.

Mill historian James Owens described the fascinating technique used to prepare the mill for its move from Hyannis Port to Orleans in an article he wrote for *Old Mill News* (October, 1983): "The entire building has been flaked.... The first step was to remove the spokes from the brakewheel and pull the windshaft. Then the cap was hoisted off in one piece and the brakewheel, pinion shaft and stones were taken out through the top." Dismantled into more than 1,000 pieces, it was trucked in sections back to Orleans. From 1983 to 1987, volunteers reconstructed parts of the windmill in an unheated airplane hangar, then moved them to Town Cove Park. The Young Windmill has since become the unofficial symbol of Orleans.

The Jonathan Young Windmill and Nantucket's Old Mill are in all likelihood the last of the 18th century windmills that remain relatively intact anywhere in the United States. According to James Owens, all machinery and parts of the Young Windmill are still in place, "except for the sweeps and tailpole, only parts of which are there."

One spring day, I drove to Hyannis Port to see where the Young Windmill had spent its years at Captain Hunt's farm. Today, the place is known as the Simmons Homestead Inn and is owned by Bill Putman. Bill proved an enthusiastic guide, with a keen interest in the history of the farm and the old mill that had been sent back to Orleans.

Postcard of the **Jonathan Young Windmill** when it was located in Hyannis Port.

At the spot where the windmill had once stood is a remarkable sight, one at the other end of the engineering spectrum from creaking wooden windmills. Putman gestured to the flagpole marking the former site of the windmill, beneath which is displayed Putman's unique collection of British sports cars. We walked among dozens of neatly arranged examples of everything from a funny little Austin Mini to a very hot Lotus—all in bright racing red.

The same scene today, with Bill Putman and his collection of British sports cars. Photo by the author.

The Morning Glory Windmill, Wellfleet, MA. Photo by the author.

The Morning Glory Windmill

145 Holbrook Avenue, Wellfleet, Cape Cod, Massachusetts

Some Cape Cod houses are accidental works of art. Wellfleet's Morning Glory house on Holbrook Avenue is an eclectic mix of an antique windmill, three mansard roofs, and a Queen Anne tower.

John Hayward wrote in his 1849 *Gazetteer of Massachusetts* that "The people of Wellfleet are engaged mostly in the coasting trade, fisheries, and the manufacture of salt; which is produced by solar evaporation." Plain windmills lined the shores, pumping up seawater for the making of salt. Hayward takes special note of the famous Wellfleet oysters: "The oyster business is also a source of revenue, furnishing employment for many vessels and men. At the first settlement of the town its bays and coves were well stored with this excellent shell-fish.... The oyster trade of Boston is principally carried on by the people of this town."

Hayward adds, "There are some manufactures of leather, boots, shoes, &c., in the town and several wind-mills for grinding corn." Wellfleet mills included one owned by David Baker on Bound Brook Island; the Thomas Higgins mill on Pamet Point Road; Freeman's Mill, north of Perch Road; and Samuel Chipman's mill near the Old Kings Highway. All of these are gone, some of their timbers having been incorporated into houses.

One of the old millstones ended up in Wellfleet Harbor, where it was dragged up by a surprised fisherman named Duffy Gardinier. Duffy used it to anchor his boat until he came upon the idea that it would make a dandy family headstone. Today, the

Gardinier name is neatly carved in the stone and it stands in Wellfleet's Pleasant Hill Cemetery, marking the graves of Lauretta B. Gardinier (March 19, 1902 – February 17, 1986) and World War I veteran Lawrence R. Gardinier (May 16, 1900 – August 2, 1973).

The mill tower attached to the Morning Glory is the only significant remnant left of Wellfleet's old smock mills. It was built by Samuel Ryder in 1838 on Mill Hill, north of Squire's Pond. Ryder used timbers from an earlier 1765 mill and ground corn there until 1870, when a Mrs. Hiller bought the mill and moved it to her house on Holbrook Avenue.

When Elton Crockett bought the Morning Glory in 1944, he converted the two privies in the yard into a workshop. Modern plumbing and other changes followed. Present owners Bill and Phyllis Crockett marvel at the number of artists who have painted the house, its armless windmill visible from Wellfleet Harbor. Streams of visitors stop and gaze at its towers, stained glass, spiral stairways, trapdoors, and weather vane with two bullet holes in it.

The Lothrop Merry Windmill, Martha's Vineyard, MA. Photo by William E. Marks.

Lothrop Merry's Tisbury Mill

Vineyard Haven Harbor, Martha's Vineyard, Massachusetts

The islands of Dukes County have a mystery and aura about them that can be sensed from a simple recitation of their names: Martha's Vineyard, Chappaquiddick, the Elizabeth Islands, and Noman's Land. They lie south of Barnstable County and Buzzards Bay. In 1602, the explorer Bartholomew Gosnold named Martha's Vineyard.

He sailed passed Gay Head, which he named Dover Cliff, and anchored in Vineyard Sound. Gosnold landed across from Martha's Vineyard at Cuttyhunk and christened it Elizabeth Island, in honor of Queen Elizabeth.

Gosnold built a storehouse at Cuttyhunk and founded a plantation that, if it had survived, would have preceded the Pilgrims by eighteen years. As Gosnold was about to leave, however, grousing and squabbling erupted among those to be left behind, and the nascent settlement broke apart. The first successful settlement by colonists was at Edgartown on Martha's Vineyard in 1641. The English crown granted William, Earl of Sterling, all of the islands between Cape Cod and the Hudson River. At first the islands were a separate colony, but in 1641, Martha's Vineyard was annexed to Massachusetts. Then, in 1664, all the islands, including Long Island and Nantucket, became part of New York for approximately thirty years. Long Island remained the property of New York; in 1692, Martha's Vineyard, Nantucket, the Elizabeth Islands, and Noman's Land became, once again, part of Massachusetts. Noman's Land, by the way, is a speck of an island below Martha's Vineyard and has the distinction of being the southernmost point of Massachusetts.

Lothrop Merry built a windmill in Tisbury somewhere between 1812 to 1815. This dome-capped mill stood on Manter Hill (then known as Mill Hill), overlooking Vineyard Haven Harbor. A barn built by Lothrop Merry's brother, Shubel, around 1820 still stands next to the Tisbury Town Hall, where it was used to store grain from the windmill. In 1833, Lothrop Merry sold the mill and barn to Captain Thomas Bradley, who in turn sold it to another captain, Tristram Luce, in about 1842. In that year, Luce moved the windmill to the north of the village, across Main Street from the Vineyard Haven Public Library. This was very near the site of a former windmill that had collapsed in 1818; Luce's father-in-law, town miller Timothy Chase, had built the earlier mill in the 18th century. The Tisbury windmill remained on this site until 1888, when General Asa B. Carey bought it and attached it to his summer home down the hill toward the harbor.

The Lothrop Merry Windmill, early 20th century, Martha's Vineyard, MA. Photo courtesy of William E. Marks.

Today, the mill and the house are visible above the breakwater which encloses the inner part of Vineyard Haven Harbor. The windmill can also be seen by looking off to the right from the ferry as it approaches

Vineyard Haven. The house itself was the home of Molly Merry from 1770 to 1843. "Aunt Molly" was the sister of miller Timothy Chase and seems to be remembered mostly as a reclusive knitter. When her husband piloted vessels out of the harbor, she would send along with him woolen stockings and other items to sell. On one of these trips, a ship was lost with all on board. It is said that when Aunt Molly heard of the tragedy she remarked, "Oh dear, all those stockings and mittens gone...."

Millstone as doorstep, at the **Lothrop Merry Windmill,** 2002, Martha's Vineyard, MA. Photo by William E. Marks.

William Marks, historian and environmentalist, wrote of the Tisbury mill in his book, *The History of Wind-Power on Martha's Vineyard*:

> *During the latter part of the 1800s, the Tisbury windmill was eventually forsakened. During this period of the windmill's gradual disuse, a Vineyard author named Hines wrote these words:*
>
> > *I did not always stand idle like this; for once the rising sun*
> > *Shone bright and gay on my long white sails*
> > *As round to their work they spun,*
> > *And I sang in joy to the favoring gales*
> > *That gave their strength till my grist was run.*
> > *But now I'm aged and gaunt, and dull must I look to the rising sun.*
>
> *Today, this windmill-house is still in existence with much of the infrastructure still intact—the grist stones now serving as steps leading into the mill.*

The Old Mill, Nantucket, MA. Photo by the author.

The Old Mill

Mill Hill Park, Corner of South Prospect and South Mill Streets, Nantucket Island, Massachusetts

No words or pictures can express the overwhelming physicality of a living, churning windmill. Sadly, it's extremely rare today to catch sight of a windmill fully set with sails turning slowly in a breeze. I approached Nantucket's Old Mill, located in Mill Hill Park, expecting to come upon a silent structure with a docent describing how it used to grind corn.

Blocked from view by a short rise, the mill didn't appear until I had climbed a few steps. Then suddenly, it burst into view, sails flapping the air, carrying its creaking old wood like a 19th century whaling ship heading out to sea. The cinematic force of the scene lacked only a full orchestra. Absurdly, what came to mind was the opening of *The Sound of Music*, with Julie Andrews rushing at me, spinning her arms like a windmill. Instead, I was greeted by the old mill's enthusiastic young miller, Patrick Prugh.

Under the aegis of the Nantucket Historical Association, there is, astonishingly, an active apprenticeship program for the Old Mill. Patrick told me, "I was made senior miller last year. This year we have a triad of millers, plus four apprentices." He had gotten his start at the notable age of fourteen, when he began hanging around the mill. Eventually, the miller asked the boy if he wanted to learn how to run a windmill.

On this day, Patrick was running the mill with two apprentices. Braced against the oak beams like a ship's pilot, he watched the speed of the blades, leaving his spot occasionally to keep the sails trimmed and to the wind, and to carefully regulate the

Patrick Prugh controlling the millstones at the **Old Mill,** Nantucket, MA. Photo by the author.

millstones. He stood holding the ropes that controlled the distance between the spinning runner stone and the stationary bedstone, all the while glancing back and forth between these stones and the mill's turning arms, which he could see through a window. He counted as the arms flicked past the window, and as the sails slowed down he moved the stones apart. When the wind picked up again and the right speed was achieved, he lowered the runner stone and the grinding continued. Patrick, as any good mariner, was alert to the weather and to every small change in the wind's strength, direction, and mood.

On the mill's second floor, apprentice Amanda Nicholas fed corn through the spout into the center of the stones, according to signals called up from below by Patrick. Apprentice Emily Chiswick-Patterson sifted the cornmeal, and sold it in small paper bags to visitors, whom she kept an eye on, lest they wander beyond the safety ropes and into the blades.

Besides Patrick Prugh, there are two other millers for the Old Mill, Garth Grimmer and Betsy Pardi. All have worked their way through apprenticeships at the Nantucket Historical Society under accomplished millers. Patrick was trained by Walter Garbalinski, and Jeremy Slavitz trained Garth and Betsy. Jeremy had, in turn, been apprenticed to Peter Kendall. Now Patrick, only a senior in high school himself, was training Emily, Amanda, Loren, and Tim.

The wind died as the afternoon wore on, and the blades slowed to a halt as if tired by their labors. Patrick went outside and stood considering the sky, and whether to

Miller Patrick Prugh with apprentices Amanda Nicholas (left), and Emily Chiswick-Patterson (right). Photo by the author.

move the tailpole and face the blades a bit more to the east. Just then, the wind picked up again and Patrick called Emily out. He went inside to hold the millstones apart as Emily grabbed onto one of the blades, threw her body into it, and then ran to start the gears turning again. The blades went part way around and stopped. While waiting for the wind to increase, I asked if I might have a try. I grabbed a blade, as Emily had done, pushed and ran. As the blade lifted into the air, I let go, spun out of the way of the following blade, and watched as the Old Mill caught the wind and gradually swept up to grinding speed.

When it was time to close up the mill and put it to bed for the night, Patrick described the process, beginning with the three methods of stopping the blades. The tailpole can be moved around to position the blades away from the wind, or the brake can be applied. The brake, however, uses the pressure of wood against wood and has the tendency to burst into flames (buckets of water are always nearby). The third way is called "choking the mill": Patrick floods the stones with corn and moves them together until they "come to a grinding halt."

Emily Chiswick-Patterson giving the blades of the **Old Mill** a running start. Photo by the author.

Next, the sails are removed from the blades, beginning with the blade closest to the ground and rotating each in turn into the low position. The sails are folded and stowed inside. The blades themselves are always left in the balanced X position. (In past eras, the set of the blades was varied to send particular signals, for example to sailing ships that needed to be signaled in cases of dangerous weather or war.) The brake is then set and, if heavy weather is expected, the blades can be chained to the tower of the mill.

Taking down the sails at the end of the day at the **Old Mill,** Nantucket. Photo by the author.

I said goodbye to the three millers, Patrick, Amanda, and Emily. Having sifted all day, Emily exhibited the most extravagant coverage of flour. As I left, she playfully used her flour-covered thumb to anoint the other two with a fair share of the dusty meal.

Nantucket's Old Mill is the last of five known windmills on the island. The stories of these mills are wreathed, if not roped and strangled, in myth. There were visions, ghosts, murder, ruin by lightning, arson, and deliberate town-sponsored explosion—giving the writing of their histories the solidity of fog.

In 1659, Thomas Macy, his wife, five children, friend Edward Starbuck, and a twelve-year-old named Isaac Coleman became the first white settlers on Nantucket. In the same year, Macy and eight other purchasers bought Nantucket Island from Thomas Mayhew for thirty pounds and two beaver hats—"one for myself [Macy] and one for my wife."

In 1665, the proprietors voted to build a gristmill near Lily Pond, to be turned—not by wind or water—but by a horse. This barely worked, so the following year it was converted to waterpower. In that year, Peter Folger, the grandfather of Benjamin Franklin, became the town miller.

A famous view painted by Philadelphia artist Thomas Birch in 1810 shows no less than four windmills on the Mill Hill skyline overlooking Nantucket Harbor. The prosperous island, one of the premiere ports of the world's whaling fleet, kept these four mills, plus the Round Top Mill on New Lane, in business concurrently.

The first Nantucket windmill was built on Mill Hill by Frederick Macy in 1723. This mill has the strangest of origins—if its history can be believed. Frederick's grandson, Obed Macy, wrote a history of Nantucket in which he related his grandfathers's

Nantucket Harbor, 1811. Woodcut by Benjamin Tanner, from *Port Folio* magazine, Philadelphia.

great carpentry skills. After building Nantucket's Straight Wharf in 1723, Frederick Macy decided the island needed a windmill, and that he was the one to build it. The only obstacle was that, supposedly, Macy had never seen a windmill. According to his grandson, "His mind became so absorbed in the subject that he dreamed how to construct the building in every part. He placed confidence in the dream and conducted the workmen accordingly. It proved a good strong mill" [8] Indeed, it proved a good mill for nearly one-hundred years. Barnabus Bunker was its last miller in 1820, when it ceased grinding corn.

The mill stood another seventeen years until the town decided to blow it up. Nantucket was struck with a devastating fire in 1836. The town looked for ways of preventing the spread of fire in the future, and hit on the idea of exploding buildings to create fire breaks. Town fathers decided to test their ability to blow things up by purchasing the old windmill. On December 7, 1837, the town gathered to watch the gunpowder ignite and the mill puff out and collapse in on itself. Less than ten years later in 1846, the island suffered the worst conflagration in its history. Fire spread quickly among the shops and houses, exploding stored oil barrels and destroying 360 buildings.

The second windmill erected on the island, the Old Mill, is the only one still standing. Said to have been built in 1746 by Nathan Wilbur, even this basic information is open to debate, for records of the exact date of construction and the name of the builder have not been found. According to historian H.B. Turner, Nathan Wilbur was a Nantucket sailor who had visited Holland and examined windmills there. He returned to Nantucket bent on building a Dutch-style mill. Islanders mocked him, but Wilbur went ahead and singlehandedly built his mill, using oak beams washed ashore from shipwrecks, and oak deck planking to sheathe the exterior.[9]

The story is a good one, but highly suspect. Turner doesn't mention Frederick Macy's 1723 windmill (a story with its own problems). And the ridicule of Wilbur's neighbors seems fabricated for effect, since by 1746 windmills were a common sight in New England and New Amsterdam/New York. Wilbur supposedly built a Dutch-style "smock" mill, the same style as stands today; but the Thomas Birch painting of 1810 shows four "post" mills on Mill Hill (post mills were boxes held above ground level by a stout, braced post). The wood for the mill may not have come from shipwrecks, for more than one source claims otherwise. For example, Edward K. Godfrey in his 1882 book, *The Island of Nantucket: What It Was and What It Is, claims,* "It was built of oak which grew just across Dead Horse Valley, to the southward of it."

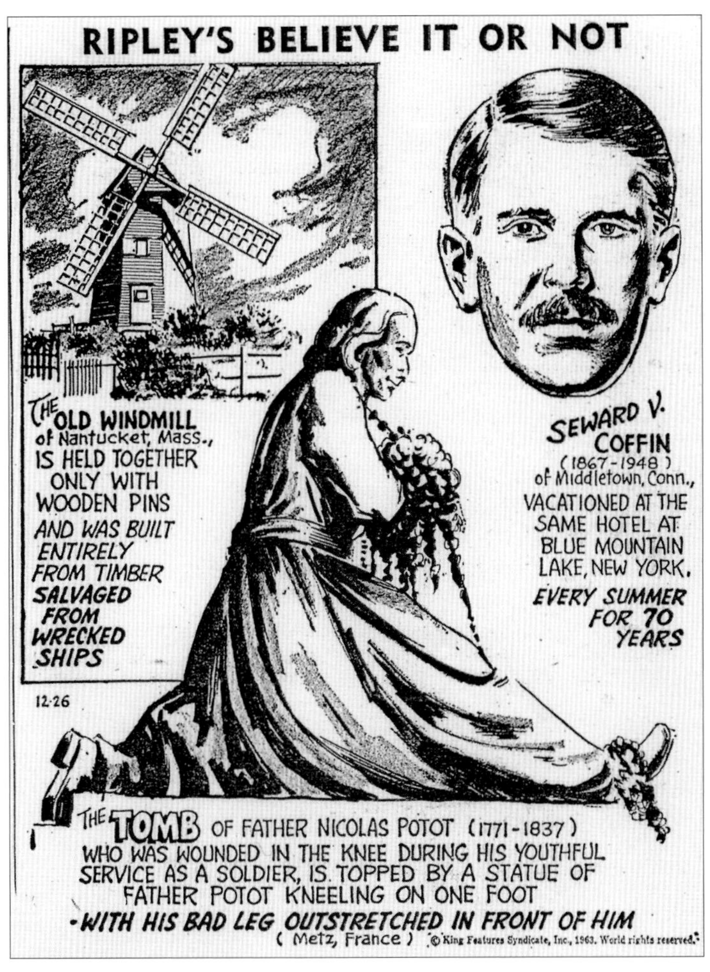

In this *Ripley's Believe It or Not* cartoon from 1963, "Ripley" chose to believe all the wood from Nantucket's Old Mill came from shipwrecks. Copyright United Media. Used by permission.

Stranger still is an undated, unsigned manuscript fragment in the Nantucket Historical Association collections, noted by NHA Editorial Board member Elizabeth Oldham. This calls into question whether Nathan Wilbur, the man credited with bringing Dutch technology to Nantucket, lived long enough to build the mill: "A company of gentlemen contracted with a man by the name of Wilbur to build the mill. Mr. Wilbur on leaving the island, with money obtained for the contract, and after reaching (the) mainland was waylaid robbed & murdered." [10]

The mill's earliest documented owners in the 1740s were Eliakim Swain and John Hay. In 1750, Timothy Swain bought it from them, only to gain the distinction of dying of a heart attack while grinding corn in the mill. After his death, Charles Swain bought it, later leaving it to his grandson, Nathan Swain. By 1828, when Nathan put the mill up for sale, it was in such poor condition that he was willing to sell it as firewood for the sum of twenty dollars.

Jared Gardner bought the mill, but Gardner, a wheelwright by trade, didn't use it to heat his shop. Instead, he refurbished the mill and put it up for sale, but Nantucket's golden age of whaling was coming to an end, and the island's sizzling economy was cooling. Gardner died of consumption (tuberculosis) in 1842. His daughter, Elizabeth Gardner Macy, her husband Peter, and George C. Gardner II, held onto the mill until 1854. It was during this period that a little girl named Caroline Dusenberry became famous for her narrow escape from the mill. Her recollection of the event can be found in the Stackpole Papers of the Nantucket Historical Association:

> *As near as I can remember, the incident took place in August of 1848. Two girls, sisters, and myself went up to the Old Mill to get some wheat to chew (that was before the days of chewing gum). The mill was not turning and one of the girls and myself climbed on one of the vanes, and the other girl pushed us, so that the vane would move a short way up and then return to its original position. We were enjoying this brief "ride" when, at that moment, the miller started the mill. The other girl promptly jumped off. But, afraid to jump, I clung to the vane. I was carried completely up and around—making a complete circle—and when I started up again I fainted and fell to the ground.*
>
> *Someone ran to my house and told those there that I was killed. My uncle carried me home and a doctor was called. I had sustained a broken thigh bone, a dislocated ankle, and bruises all over my body.*[11]

In 1854, the mill was sold to George Enas at the rock-bottom price of $150. Enas was the first of three Portuguese owners of the Old Mill. He had been born in 1815 on Flores, the westernmost island of the Portuguese Azores. In 1866, Enas died of consumption (as Jared Gardner had before him), a lung condition aggravated perhaps by years of breathing the flour and dust of the mill. Enas, only fifty years old, must have known he was going to die, for a year before, in April 1865, he had sold the mill to Captain John Murray, from Graciosa, another island of the Azores. According to Helen Seager's article, "Portuguese Islanders and the Old Mill," (*Historic Nantucket*, Spring 2002), the sale included "sixteen picks, two jacks and falls, one crowbar, one handspike, one capstan, and the measures and fixtures. Also the goodwill of the trade, the said Enas hereby agreeing not to carry on the business of milling on Nantucket."

Murray kept the mill only twenty months, selling it at a large profit in December 1866. Later, he set off for a five-month whaling voyage to the Azores. The buyer, John Francis Silva (later recorded as "Silvia") and his wife, Frances, arrived on Nantucket from the Azores in the 1860s.

The Portuguese arrived on Nantucket at a time when the booming whaling economy was well past its peak. By the 1820s, New Bedford had eclipsed Nantucket in whaling. The once great Nantucket whaling fleet contained only sixteen ships by 1846, and was further reduced to four by 1857.

The 1850s saw an exodus of Nantucketers, a population that was by that time a rich mix of cultures. The neighborhood surrounding Mill Hill was, in fact, known as "New Guinea." In the early 1800s, ex-slaves and African-American mariners had settled around the four windmills. Both whites and blacks left for better prospects in the 1850s, some heading to the gold fields of the West. Concurrently, the Portuguese saw opportunities for themselves on Nantucket and migrated to the island in great numbers. The newcomers were able to buy land and businesses at falling prices.

Engraving of the **Old Mill,** *Harper's New Monthly Magazine*, 1875.

The first suggestion that the Old Mill was worth preserving as an historic site came in 1877, after its blades were badly damaged in an autumn gale. Fearing the mill might close down, a letter to the editor of the *Inquirer and Mirror* suggested that "the mill could be made a source of profit" by placing "an observatory in its top [that] would bring in many dimes" and "a small restaurant on one of it upper floors...a few windows [could] be put in the dining room, and a hungry crowd would enjoy...sitting...among the cobwebs and dusty beams." [12]

Silva repaired the mill and continued grinding corn, but the future of the mill still seemed shaky. In July of 1892, the following notice appeared in the *Inquirer and Mirror*:

A rumor was current early the present week that the old mill had been sold to a gentleman representing the Columbian Exhibition and was to be taken to Chicago and be re-built. A feeling of regret not unmingled with indignation was aroused throughout the community at the thought of removing this ancient landmark, but we are glad to be able to state that the rumor is unfounded. The Old Mill is to our visitors one of the principal objects of interest on the island—the Mecca of strangers and pride of our own citizens. Within and around it hallowed memories cluster. It has withstood the storms of a century and a half unshaken. Let no ruthless hand disturb it. Hand over the antique furniture of our ancestors to the bric-a-brac hunter if we will, but spare the Old Mill.

When John Francis Silva died in April 1896, the Old Mill was left to his heirs in the Azores. In July 1897, an ad appeared in the *Inquirer and Mirror* putting the mill up at auction. The Nantucket Historical Association had already been raising funds for the possible purchase of the mill, and the Association bid $885, which was $135 more than it had raised. Summer resident Caroline French made up the difference.

It would seem the future of the mill was secure, but the 20th century proved unkind, beginning with a severe gale which damaged it in 1913. The exterior was repaired and the opportunity taken to overhaul the machinery so that corn could be ground for the first time since 1892. This lasted only until 1928, when the machinery again deteriorated and grinding stopped. The mill suffered more storm damage in 1932, but was completely restored in 1936.

Catastrophe arrived in the form of the Great Blizzard of February, 1952. The furious storm swept across Nantucket and slammed into the mill, breaking its main shaft. Wet snow clung to the blades like loose cement until the rushing wind caused the blades and the shaft to come crashing down in a heap. This was described in the local press as "no single, crippling blow; this was a complete disaster." Within six weeks, a piece of white oak was located in Portland, Maine, for the main shaft, and white ash and spruce were bought for the blades. A large crane was brought in to reassemble the mill the following August.

The Old Mill was next threatened deliberately, by vandals. The January, 1970, issue of *Historic Nantucket* reported, "In the pre-dawn darkness on Friday, August 21, attempts by persons as yet unknown to destroy two of Nantucket's historic landmarks by fire came frighteningly close to accomplishing this grim purpose." The Old Mill and the Old Jail had been simultaneously set on fire by arsonists. The article continues:

These unprecedented incidents sent a chilling wave of horror and repulsion throughout the community... The saving of the Old Mill was a near miracle. Two young policemen—Officers Hunt and Holloway—were driving past the structure on regular patrol when the latter spotted the flames and called headquarters from the cruiser.... At the Old Mill's north side a dangerously large area of the shingles had caught fire, apparently spread by the use of kerosene, and had not the apparatus arrived so promptly it was evident that the flames would have soon entered the structure.

Another policeman, Ed Dougan, had more pleasant, though eerie, memories of the mill during the 1970s. Dougan was the miller from 1977 to 1980, and recounted to writer Blue Balliett (in her book, *Nantucket Hauntings*) his experience with ghosts. Dougan recalled the death of miller Timothy Swain in the 18th century:

Timothy Swain was an eccentric man. He was either a devoted miller or a poor sleeper, for when the winds were favorable he would run the mill at night. People would say that Swain was out doing a "nocturnal twirl." One morning the town woke up to find the mill grinding away at dawn and Timothy Swain dead inside. He had died, apparently, of natural causes.

Dougan found that when he got the mill running at the right speed, he could tie off the beam and let it run on its own—but then something odd happened:

As soon as I got to the other side of the room, directly across from that beam, the mill would suddenly pick up speed and I'd have to go running back. It seemed that I could make this happen. The mill would be running in a good steady wind, without needing any adjustment, and I'd walk exactly to that halfway point by the stairs, and the mill would take over. It happened literally hundreds of times.... I could do it in front of a crowd of people. I'd show them where I was going to walk and what was going to happen when I got over there, and it always happened just as I'd predicted.

When I ran into those inexplicable shenanigans in the mill.... I immediately thought of Timothy Swain. Of course, there's no way to prove who or what was responsible, but anyone who was in that building grinding corn by moonlight had to know what he was doing. A nocturnal twirl would be a job for an experienced miller. Timothy Swain owned that mill and died in it. It makes sense that he would still have a finger on the works.

Prescott Farm Windmill

2009 West Main Road, Middletown, Rhode Island

When I saw his hat stuck with duck feathers like a pin cushion, I knew there was something unusual about John Lingley. John has been caretaker of the Prescott Farm Windmill on Aquidneck Island for fourteen years. He sat at the old wooden counter of the Country Store which serves as the Prescott Farm's reception office. Langley looked very much the part of the storekeeper waiting for his next customer. Except for the hat.

The flock of ducks I could see through the store's rippled glass window seemed to provide a clue to John's odd headgear. He explained that when he takes school children

Woodcut of the Nantucket windmill on bag of meal ground at the mill, 2002.

around the grounds of Prescott Farm, he's soon followed by the farm's population of white ducks. Many a student has drawn a picture of John wearing his hat covered with feathers, and surrounded by ducks with the windmill in the background. For years the children of Aquidneck Island have sent their pictures addressed to "The Duck Man," and the name has stuck.

Prescott Farm in Middletown is really a museum collection of 18th century buildings, with a windmill at one end and a duck pond at the other. The Newport Restoration Foundation bought the Henry Overing farm and restored the farmhouse (now known as the Prescott House), which was built about 1730. The John Earle House (built about 1715), and the windmill (circa 1812) were both moved to the site from other locations. John Earle built his house at Bristol Ferry Road in Portsmouth, where it served as the ferry-master's house. Today, it is used as a country store where visitors can buy honey, medicinal and culinary herbs, edible flowers, and cornmeal.

The striking smock windmill with a dome cap was built in Warren, Rhode Island, around 1812. One of the great old millers of Rhode Island, Benjamin F. C. Boyd,

John Lingley feeding the ducks at the **Prescott Farm Windmill,** Middletown, RI. Photo by the author.

recalled an unusual use of this mill—its role in the distilling of whiskey at its Warren location. (Whiskey was a very popular drink in early New England; great kegs of it were broken open to celebrate the raising of houses, barns, and even churches). The windmill was moved to Fall River, Massachusetts, then to Quaker Hill in Portsmouth, Rhode Island, by Robert Sherman. Benjamin Boyd, born in 1862, recalled, "When I was about 14 years of age it was bought by the late Benjamin and William Hall, and moved to Lehigh Hill on the West Road."

In the early 20th century, the windmill was converted to gasoline power. Then, after lying idle for many years, it was bought by the Newport Historical Society in 1929. The Society restored it to wind power, then restored it again after it was damaged in the Hurricane of 1938. In 1968, Doris Duke and her Newport Restoration Foundation purchased the windmill, and in 1970 moved it in sections five miles down West Main Road to Prescott Farm.

After taking a couple of photos of the Duck Man, his flock and the windmill, we returned to the Country Store, leaving my wife, Karen, with the ducks. While John was telling me more about the farm, Karen came rushing in saying something was wrong—one of ducks had its legs bent and a wing was stuck out at an odd angle. The Duck Man looked out the window and said, "Oh him! He's fine, he just had too much 'female companionship' over the weekend." As we left Prescott Farm in search of more Rhode Island windmills, the Duck Man presented us with a pound of stone-ground cornmeal in a sack illustrated with the old Prescott Farm Windmill.

The Wyatt Windmill

140 Smith Neck Road, South Dartmouth, Massachusetts

The Middletown, Rhode Island, mill with the strangest history is known as the Wyatt Windmill. Built in Rhode Island, it was moved to Massachusetts twice, the second time by Colonel E.H.R. Green, only son of the woman known as "The Witch of Wall Street." According to the Middletown Historical Society, the mill was probably built in Bristol, Rhode Island,

The Wyatt Windmill, Middletown, RI, 1890. Photo courtesy of the Middletown Historical Society.

in the middle of the 18th century. Its first move was by water—or ice—to the area now known as Fall River, Massachusetts. It was later moved back to Tiverton, Rhode Island, and then to Newport. From Newport it traveled to Middletown, near the junction of Mitchell's Lane and East Main Road.

By 1797, it was known as the Chase Windmill. When Jethro F. Mitchell owned the mill, the *Newport Mercury* of January 3, 1829, noted, "Jethro F. Mitchell mill robbed. Bags of meal, barley meal [taken], Middletown." And the Newport *Herald of Times*, March 1, 1849, advertised, "Windmill at auction, Jethro F. Mitchell, Middletown."

Around 1855, Samuel Wyatt bought the mill and he and his son, Nathaniel, moved it to Wyatt Road, near today's Methodist Church. Robert Wyatt inherited the mill upon his father Nathaniel's death. Of Robert, William Hallenbeck (owner of Wyatt's mill as of August, 1981) wrote:

> *Robert Wyatt was a talented miller, and under his ownership "Wyatt's meal" was in great demand, because of its superior quality. He was diligent in "picking" his grindstones monthly, and certainly was adept in turning the arms off the wind if the speed of the runner stone was excessive. Robert Wyatt cared about his windmill, which was oak-hewed, braced across and braced between. During his lifetime, he placed spruce three-by-fours all over the timbers that were worn to thoroughly reinforce them.*[13]

After Wyatt's death in 1912, Arthur Brigham ran the mill until 1923, by which time flour and meal from grainfields in Midwestern states made mills in the East unprofitable. Then things got really interesting for the Wyatt Windmill....

Hetty Green had made enormous sums of money in the stock market in the early part of the 20th century. Described as both ruthless and miserly, she was given the unfortunate title, "The Witch of Wall Street." After her death, her only son and heir, Colonel E.H.R. Green, inherited her fortune, as well as her summer home, Round Hill, in South Dartmouth, Massachusetts.

The Wyatt Windmill, Middletown, RI, with bags of meal inside the door, 1890. Photo courtesy of the Middletown Historical Society.

Green had an active mind, an eccentric imagination, and plenty of money. In 1919, he began building a mansion at Round Hill, where he moved with his wife, an ex-stripper. Green established a hot air balloon station at Round Hill, lost interest in it, and next discovered windmills. He found the Wyatt Mill near Newport, a Green family haunt, and moved it to his South Dartmouth mansion in 1924 to serve as a landscape ornament.

Colonel Green went on to build the first radio tower in South Dartmouth. His interest in aviation led him to blimps and airplanes, which led to meteorological studies, then to wave studies, and finally, nuclear studies. Green's most lasting legacy did not come from his dabbling in science, but from his hobby of collecting historical objects. Foremost was his purchase of the *Charles W. Morgan* whaling ship, the last whaler to sail out of New Bedford. He kept the ship in dry dock at Round Hill estate. When Colonel Green died on June 8, 1936, he left little money for the care of the *Charles W. Morgan*, or anything else. The Hurricane of 1938 further damaged the vessel. Fortunately, Mystic Seaport Museum rescued the ship and moved it to Connecticut for restoration in 1941.

As for the Wyatt Windmill, it sat neglected, and, like the *Morgan*, suffered damage in the Hurricane of 1938. Fortunately, it was reclaimed by the Hallenbeck family. Today, it sits in fine condition where Colonel Green left it, behind what are now the high hedges of the Hallenbeck house on Smith Neck Road, South Dartmouth.

Boyd's Windmill

Middletown Historical Society, Paradise Valley Park, Middletown, Rhode Island

When Benjamin F. C. Boyd turned eighty years old in 1942, he looked back on his life as the last miller of the famous Boyd's Windmill. Boyd epitomized the classic New

Boyd's Windmill, Middletown, RI, 2002. Photo by the author.

England windmiller—tough, thoughtful, something of an inventor, a politician, and a poet. His stunning eight-vaned windmill is the centerpiece of the Middletown Historical Society's programs.

Boyd's Windmill began its long life in Portsmouth, near the corner of West Main Rd. and Mill Lane. The best teller of his tale is Boyd himself, who published his reminiscences in the *Fall River Herald News* in February, 1942:

> *I do not know which is the oldest mill now standing, but as the one I own and run, and which is known as Boyd's Mill, is quite old and as my family has probably had a longer experience in running wind grist mills than any other family outside of Europe, it may be of interest to describe our mill and our connection with it.*
>
> *It was built in the year 1810 by John Peterson, a retired sea captain; the timber being cut back of Wickford Village on the west side of Narragansett Bay....*
>
> *Few of those now living know what a tough job it is to run one of these old wind grist mills, especially in the Winter. There were no stoves in them because owing to the revolving top there could be no chimney, and owing to the shape of the mill and its windy position a stove pipe run out the side of the mill would meet 57 varieties of draft, under which no fire could burn; also one door had to be kept open so that the miller could keep his weather eye peeled on the weather, lest he be caught and wrecked in a squall. In fact, the running of the wind grist mill required about the same ability and judgement as a sea captain, other than he did not have the water to contend with....*
>
> *It was the miller's business to make a fine quality of meal, but as his power and speed constantly varied in a gusty wind he had to constantly vary the amount of grain flowing between the stones and the weight of the running stone on the grain, and in the midst of it all some farmer's horse would become frightened at the mill and start to run away. Then the miller would have to leave off what he was doing and find the farmer's grist for him and put it in his wagon, while the farmer held his horse and tried to take a reef in his broken harness and wagon.*
>
> *Many are the days that I have run this mill when the thermometer was down to zero, in a howling gale and snow storm, when one could not see 200 feet; but the grinding had to be done, and these millers under such conditions had to use all their experience and nerve to turn the power of the wind into ground feed for man and beast.*
>
> *When the miller was running his mill on his nerve in one of these storms, he could not tell whether in five minutes a heavier squall would wreck him, or whether he might be in the midst of a calm. Another hazard of running these mills was in trying to do so in a near freezing rain storm. In 10 minutes the sails might freeze as stiff as a board, then he would be at the mercy of the wind as he could not furl his sails....*
>
> *When we thought it was going to rain in the Winter we took the sails off, but*

we sometimes left them on and they got wet and froze up. I have spent many hours up on those mill arms with the temperatures down to zero picking out frozen mill sails, or trying to thaw them out with teakettles of boiling water from the kitchen stove.

What I have written here is history. I know wind grist mills from the ground up. When I was four years old I was struck on the head by one of the mill arms and picked up for dead; and when I was 28 years of age I went up on an arm to fix a sail. I did not put the brake on but braced the arm against the ground with a stick, the stick jarred out and the mill started, and I went sailing through the air, around and around. I barely managed to twist my arms and legs around the slats. I had to hang on in all positions, and every time I went over the top of the circle I was head down 45 feet from the ground. My brother Edward was up in one of the fields digging out rocks. I hollered to him and saw him start on the run, then I closed my eyes and paid all my attention to hanging on and tried to ignore as to whether I was head down or sideways. It was a matter of keeping my head, and holding 150 pound of dead weight to that slat work until my brother got there.

It was estimated I went around 30 times. My brain has been in a whirl ever since, cause and effect.

Boyd's grandfather, William, had owned another windmill, near Bristol Ferry, which was destroyed in the Gale of 1815. William Boyd leased, then bought, John Peterson's Portsmouth mill, beginning nearly 130 years of family ownership. In 1855, when Benjamin Boyd was 23 years old, he rebuilt the mill, replacing much of its inner workings. Boyd continues his story:

All meal up to 1895, whether sold in the cities or farmers' grists, was unsifted and the house wife had to do it. At this time I invented a power sifter run by the mill.... I have made all the repairs on this mill since I was 18 years of age. I know how to build one of these mills, and in 1901 remodeled our mill from a four armed mill to an eight armed mill. It was the wonder of all who saw it, and there never was one like it stood on the earth since Adam was placed in the Garden of Eden. It was an experiment like Noah's Ark.

Postcard of **Boyd's Windmill.** Courtesy of the Middletown Historical Society, RI.

Boyd's Windmill delivery truck, 1930. Photo courtesy of the Middletown Historical Society, RI.

It is doubtful if there will ever be another of these wind grist mills built in this country. The timbers and boarding in the lower stories of these old mills are full of rusty tack nails, where notices of town meetings, auctions, and other notices have been tacked up, and the old blacksmith shops were used the same way. They were the public forums where news, politics and scandal were debated....

I can remember when there were 10 of these mills standing in Portsmouth, Middletown and Newport, two in Little Compton, and before my time there was one in East Greenwich, and another in Bristol or Warren, but old Father Time came out ahead in the race and only a few are now left.

I have heard my father say that in the Revolution there was one of these windmills stood on Prudence Island, and the British soldiers used it as a target to try their cannon on. To the people here on the Island these old mills are so common-place that they excite very little interest, but to the thoughtful person, or stranger, they are always a source of wonder and curiosity, possessing for him an air of mystery and historical interest, as their moss covered shingles tell him that these old mills form a connecting link between him and his ancestors....

Boyd's Windmill, West Main Street, Middletown. Early 20th century. Photo courtesy of the Middletown Historical Society, RI.

On Benjamin Boyd's 80th birthday, the *Fall River Herald News* recalled that he had served in the House of Representatives from 1918 to 1926, where "he scored victorious points over political opponents with snappy verse served up with a healthy share of his homey philosophy." Of his verse, Boyd said, "It's just facts put to doggerel verse. I know no more about poetry than a pig does about ice skating." Here is Boyd's take on a familiar saying:

The mills of the gods grind slowly
But they grind exceedingly small.
While man's wind mills and ambitions vanish
And Death takes the miller as toll.
All is vanity and vexation of spirit.

I visited the Middletown Historical Society on a fine autumn day in September when the fifth graders from the local Gaudet Middle School were about to tour Boyd's Windmill. They were there for the Society's popular "Wonderful Old Windmill" program.

I entered the headquarters and museum of the Society, housed in Paradise School, one of two one-room schoolhouses the Society has restored. Their delightful archivist, Mary Bellagamba, showed me detailed models of the windmill and the Society's collection of vintage windmill photographs. While Society president Stanley Grossman waited for the next class of fifth graders to arrive, he showed me around the museum exhibits and told me more of the windmill's recent history.

Fifth grade students from Gaudet Middle School, Middletown, RI, enjoy a windmill program by Stanley Grossman, President of the Middletown Historical Society. Photo by the author.

By 1990, when the Boyd heirs donated the mill to the Middletown Historical Society, it had been idle for nearly 50 years and its arms were gone. Mr. Grossman told me that this was the only known eight-vaned windmill in the United States, but there are others in Europe. Benjamin Boyd's hope in 1901 that eight vanes would be better than four was not entirely successful. As Grossman commented, "Boyd was not a very good engineer. It fell apart." In 1915, Boyd converted the mill from wind power to gasoline power. Before rebuilding the vanes, the Society wisely sent their contractor to England to study the mechanics of an eight-vaned windmill.

A restoration team was assembled under millwright restoration specialist Andrew Shrake from Dennis, Massachusetts, and architect Richard R. Long of Newport, Rhode Island. Looking for creative modern ways of moving a windmill, they found a helicopter company that agreed to do it—for $100,000 a day. Instead, the more traditional method of dismantling and numbering the pieces was employed. The windmill parts were trucked from Mill Lane in Portsmouth to Paradise Park in

Middletown on August 1, 1995. It took four more years to restore the mill to working condition, complete with its unusual array of eight sail-covered vanes.

The dedication of the restored **Boyd's Windmill,** October 21, 2001. Photo courtesy of the Middletown Historical Society, RI.

I approached the windmill along a winding path in Paradise Park. Set in an open field, with the church spires of Newport visible on the horizon, Boyd's Windmill, with a few cows added, could easily be imagined as a John Constable painting. On this particular day, children, instead of cows, surrounded the mill, some sketching it, others gathering around volunteer Bill Miller for a tour. When I joined the group, Bill was explaining the chain wheel on the side of the windmill's dome. From the wheel, a chain extends to the ground. By pulling on the chain, the miller would turn the top of the mill so the sails faced into the wind.

Inside the mill, Bill pointed out the pegging of the beams, without a single nail to be seen. He described how hard-flint corn was grown and then dried in corncribs for at least eight months, and sometimes up to two years. After shelling and winnowing, the corn was ready for grinding. The students peppered Bill with questions—some relevant, some just curious. (One student wondered if Bill's missing finger had anything to do with the gigantic granite millstones.) Their excitement was genuine when the class was presented with a sack of cornmeal and recipes to make jonnycakes. It seemed to me that Benjamin Boyd, the windmill's last miller, would be deeply satisfied by this 21st century scene. As Boyd said, in his reminiscences,

> *When I think of the changes that have taken place since my grandfather ran this same mill, and to which it has been a mute witness, it seems as if the Great Power*

which causes these ceaseless changes in the universe should endow it with the power of speech, that it might speak in counsel to the grandchildren of those it served so faithfully long years ago.[14]

The Newport Tower, Touro Park, Newport, RI. Photo by the Author.

The Newport Tower

Touro Park, Corner of Bellevue Avenue and Mill Street, Newport, Rhode Island

The most mysterious windmill in New England—and the most controversial—is the Newport Tower in Touro Park. In 1854, *Harper's New Monthly Magazine* called it the "one monument which interests the poet, the antiquarian, the traveler, the controversialist, the divine; of which sweet songs have been sung, wild theories spun, and happy hoaxes invented." Built of stone, it has eight open arches separated by columns, the first and fifth of which are aligned in a North-South configuration oriented by the North Star.

For more than 250 years, scholars and poets have argued over the tower's origin. Was it a pagan temple, church, fort, gunpowder house, watchtower, lighthouse, windmill, or a combination of these? Was it built by the Druids, Phoenicians, Romans, Vikings, medieval explorers, Native American sachems, Rhode Island's first governor, a Portsmouth mason, or a Portuguese sailor?

Historians like Dr. Manuel Luciano DaSilva have referred to the Newport Tower as "the single most enigmatic and puzzling structure to be found in the United States," a building "which may hold the key to an unknown portion of the history of the Western Hemisphere." [15] Only recently, at the very end of the 20th century, has the modern scientific technique of Carbon 14 dating solved the mystery of the Newport Tower.

As early as 1839, rumors were afloat that the tower had been built by Vikings. According to scholar and theologian Dr. William Ellery Channing, the Society of Danish Antiquaries at Copenhagen had studied drawings of the tower and decided that it was probably built in the 12th century by Norsemen. As the theory of a Viking origin of the Newport Tower developed, the Danes attributed it to Bishop Eric Gnupsson, who, they believed, built the tower as a Christian church in the 12th century. A rumor of another Viking discovery spread in the 20th century. It purported that the remains of a Viking ship were uncovered just after the Civil War by workmen building Newport's Ocean Drive. The Newport Historical Society has no evidence of the discovery.

In 1942, writer Philip Ainsworth Means, in a book about the Newport Tower, threw his support to this Viking theory, though he allowed that "any ordinarily prosaic person might be excused for regarding it as fanciful balderdash." [16]

Henry Wadsworth Longfellow fueled the Viking legend of the Tower with his 1840 ballad, "The Skeleton in Armor":

Three weeks we westward bore,
And, when the storm was o'er
Cloud-like we saw the shore,
Stretching to leeward;
There, for my lady's bower,
Built I the lofty tower,
Which, to this very hour,
Stands, looking seaward.

There lived we many years;
Time dried the maiden's tears;
She had forgot her fears,
She was a mother;
Death closed her mild blue eyes,
Under that tower she lies;
Ne'er shall the sun arise
On such another!

Two years earlier, Longfellow had seen an actual skeleton dressed in armor, which had been dug up in Fall River, Massachusetts. He supposed it to be "the remains of one of the old Northern sea-rovers, who came to this country in the tenth century." Longfellow continued to believe in the Viking theory, later reading about it in a series of letters at Brown University. The letters were dated 1847 and signed "Antiquarian." But they proved to be a hoax. As *Harper's New Monthly Magazine* reported in August, 1854, the letters included "fabulous investigations by fictitious characters, which did not fail of provoking caustic correspondence, and finally achieving its triumph by eliciting a solemn denial, from

Artist's concept of the **Newport Tower** when equipped as a windmill. Courtesy of the Middletown Historical Society, RI.

Professor Rafn, of the Royal Society of Northern Antiquaries at Copenhagen."

The first documented mention of the stone tower occurs in the deed for the Jewish Cemetery, Beth Chayim, dated February, 1677. In it, Newport cooper Nathaniel Dickens sells a property bounded by "ye Stone Mill," and authorizes the land for the use of the "Jews and their Nation Society or Friends." Within a year, the mill is mentioned again, this time in the will of Governor Benedict Arnold, dated December 20, 1677. Arnold, no relation to the famous traitor of the same name, came to Newport from Providence in 1653 after some difficulties with Roger Williams. The eastern part of Arnold's homestead included what is now Touro Park. In his will, Arnold speaks of the lot upon which stands "my stone-built wind-mill."

Significantly, neither Arnold, nor any explorer, settler, or Native American left any trace of having noticed the stone tower before 1677. In 1663, one of the area's first settlers, Peter Easton, wrote in his journal that the first windmill in Newport was built in that year. In 1675, it was blown down by a heavy gale. Governor Arnold's stone windmill may have been built to replace this one. (The Viking theorists claim he could have built it on the existing Viking stone tower.) Others note the striking similarity between the Newport Tower and the Chesterton Windmill in Warwickshire, England. It is supposed that Arnold was born in Ilchester, Somerset, about 100 miles from the Chesterton windmill. It is possible he may have seen the Chesterton mill, also built of stone with eight arches.

A stone windmill in Chesterton, England, similar to the Newport Tower. Photo courtesy of the Middletown Historical Society, RI.

Until about 1755, the Newport Tower windmill ground corn. Then, for nearly ten years, it was used as either a powder mill or magazine. In 1764, it began service as a haymow, and was painted as such by Gilbert Stuart between 1770 and 1775. For the next five years, the British used the tower as an ammunition magazine. On leaving Newport, they tried in vain to blow up the building, but they succeeded only in blowing off the tower's top two or three feet.

Governor George Gibbs, Jr., acquired the property in 1799. Fifty-five years later, Touro Park was founded as the gift of Judah Touro, with the tower as its centerpiece. The Old Stone Mill, or Newport Tower, became a cherished, though enigmatic, part of the Newport social scene.

In 1948 and 1949, archeologists sponsored by The Preservation Society of Newport County made several digs at the tower. They exhumed bits of glass, pottery, and clay pipe stems, but nothing of Viking origin. Then, in January of 1993, samples of the lime mortar from deep in the tower walls were removed for the purpose of Carbon 14 dating. These were sent to Helsinki University's Institute of Physics and Astronomy. On September 22, 1993, the Committee for Research on Norse Activities announced the results in Newport: According to Carbon 14 dating, the tower was not built before the arrival of Columbus, and was almost certainly built in the 17th century. Newport seems undaunted by this turn of events, and some are holding out for a piece of Viking bronze or a horned helmet to turn up from under the stone tower.

I asked the proprietor of the Sea Whale Motel, where I was staying, what he thought of the supposed Viking origins of the tower. Bob Henninger, bearded like a Viking himself, seemed sorry to see the legend debunked and offered to pose in front of the tower in a Viking helmet. With a wry smile, he said the "Greek Vikings" (owners of various "Viking" pizza parlors) would be mighty disappointed. Meanwhile, Newport's three-figure telephone code still represents "VIKing", the local football team continues to be called the Vikings, and on the streets can be found the likes of Viking buses, the Hotel Viking, Viking Insurance, the Viking Uniform Company, and Viking Cleaners.

Early 20th century postcard of the **Newport Tower** in Touro Park.

Windmill Cottage

144 Division Street, East Greenwich, Rhode Island

Henry Wadsworth Longfellow backed the romantic Viking tower theory for the Newport Tower, and was equally taken by the romance of authentic old windmills. In 1870, in fact, Longfellow bought an abandoned windmill in East Greenwich on the coast of Rhode Island midway between Providence and Newport.

Longfellow had an enduring friendship with Professor George Washington Greene, whose 18th century home in East Greenwich had, unfortunately, passed out of his family. As an exceptionally generous sign of their friendship, Longfellow bought the home as a gift for Greene and his wife. Four years later, Longfellow bought the old windmill on the southeast corner of Division and East Streets, and had it moved to the Greene house on the southwest corner of the same intersection.

The four-story, eight-sided smock mill with typical Rhode Island domed cap still stands connected to the house at 144 Division Street. Professor Greene used the first floor as his study. Longfellow, a frequent visitor, slept on the second floor of the mill. In 1880, Longfellow sent his friend Greene a poem about the windmill, which begins:

Behold a giant am I
Aloft here in my tower
With my granite jaws I devour
The maize and the wheat and the rye
And grind them into flour....
From "The Windmill"

The Jamestown Windmill

North Road, Jamestown, Rhode Island

The most satisfying approach to the Jamestown Windmill on Conanicut Island is from the Newport Bridge. Leaving Newport, one crosses the swooping, nearly two-mile-long bridge, with Gould Island in Newport Bay to the right, and Rose and Goat Islands, each with a lighthouse, on the left. Making the steep descent at the Jamestown end of Newport Bridge, one sees the rolling fields of Conanicut Island. There, among an English countryside of meadows and stone walls, can be

The Jamestown Windmill, RI. Photo by the author.

seen the dazzling sight of the Jamestown Windmill.

Located to the north of the Friend's Meeting House at the junction of Weeden Lane and the North Road, the Jamestown Windmill is a good example of the triumph of free enterprise over the inefficiency, and sometimes abuse, of government-funded projects. In 1787, Jamestown decided to build a town windmill and hire a miller to run it. For land, the town asked the state for a half-acre piece of Col. Joseph Wanton's farm. Wanton had been a Tory in the American Revolution, and had left the island along with the British when they evacuated Newport. All of Wanton's property had been confiscated by the state.

To finance the mill, Jamestown sold off some of the island's roads in exchange for 100 silver-milled dollars. Jethro Briggs was hired as its first miller and was loaned the 100 dollars as capital to begin the business of grinding the island's grain. Briggs was to pay back the loan in one year with 100 dollars worth of Indian corn, and to pay the town 200 bushels of "good merchantable Indian corn" when the lease was renewed (as it would be, if he kept the mill in good repair). Briggs' salary, called a "toll," was to be three quarts of flour or meal, which he could keep out of each bushel he ground for customers.

Within six months, it was clear that the arrangement was a disaster. Complaints against Briggs included his frequent absences from the mill, overcharging on the toll, producing poor quality meal, and failing to pay back his loan. It took years to clear up the messy financial arrangement between Jethro Briggs and Jamestown, at the end of which, in 1795, he was let go and the mill was sold to Benjamin Carr, a private buyer. Carr backed out within months, and by the end of August it was re-sold to Nathan Munro. Munro proved the mill could be operated profitably, and did so for thirty-one years until he retired.

For 109 years, the Jamestown Windmill ground cornmeal and animal feed until the big new rolling mills of the Midwest put it out of business. Abby Tefft was its last owner, and her brother-in-law, Jesse, was the last Jamestown miller to grind meal for the famous Rhode Island Jonnycakes. The Jamestown Historical Society, which acquired the windmill in 1912, provides the following recipe:

Rhode Island Jonnycake
Mix one cup Rhode Island cornmeal, 1 tsp. salt, 1 tsp. sugar. Add one and a quarter cups of boiling water and mix well, then thin with about a quarter to a half-cup of milk. Drop by tablespoonsful onto a well-greased griddle or heavy frying pan over medium heat. Cook for 6 minutes, turn over and cook for about 5 minutes. Serve with butter and syrup.

The windmill had stood derelict until a number of ladies decided to save it around 1904. Mrs. Frank H. Rosengarten of Philadelphia kicked off the fundraising by giving a whist party at her house. Serious restoration began after the founding of the

Historical Society in 1912. There were problems, however, as the minutes of the Society show:

> *June 17, 1916. The secretary was authorized to notify Mrs. Boone that unless she kept her cows out of the Windmill yard they would be placed in the pound.*

The Jamestown Windmill has been battered and broken by hurricanes, restored and restored again. A major restoration, at great cost, was completed in 1970. Thousands arrived for the celebration. Then a freak gale hit in March, 1974, snapping off all four vanes at the windshaft. The latest restoration occurred in 2002, with peripatetic millwright Andy Shrake in charge. ●

Chapter Four

The Lost Windmills

"Man, in his effort to harness the forces of nature, has produced two beautiful wind drinking contrivances—the windmill and the sailing ship. There is a beauty, individual and uncounterfeitable, in every sailing ship and windmill, a charm direct and personal as the charm of a friend.

"..A sailing ship traversing the ocean is to me as wonderful and mysterious as a meteor crossing the heavens, and windmill sails revolving against the blue and green of quiet lands arouse in me feelings as deep and mystical as those with which I regard the remote and whirling stars."

Frank Brangwyn & Hayter Preston, *Windmills*

The 1787 **West Falmouth Windmill** was moved to Brockton, MA, to house an ice cream shop. It burned down in 1922.

The West Falmouth Windmill

West Falmouth, Cape Cod, Massachusetts

The story of the West Falmouth Windmill is even sadder than that of the Farris Mill—the old West Yarmouth windmill relocated to the Henry Ford Museum in

Dearborn, Michigan. Vintage Cape Cod postcards show a large, very handsome windmill with a sign proclaiming, "The Oldest Wind Mill on Cape Cod." To the town's dismay, however, the famous Eastham Windmill proved to be even older. The final insult came in 1922, when the old West Falmouth Windmill ended its days as a pile of black ashes in Brockton, Massachusetts.

As early as 1659, settlers from Barnstable began eyeing the land Native Americans called Succonessett. In 1686, Falmouth was officially incorporated and soon became a major whaling port, ship builder, and manufacturer of woolens. Unlike most other Cape towns, Falmouth had good waterpower, and the Dexter water mill was built on the Coonamessett River in 1700. However, town records show that free enterprise had its bumpy moments in Falmouth:

> *In 1719, much complaint was made of the miller. The town had previously some difficulty with this Mr. Philip Dexter; but, as he had no competitor, and the people were dependent on his mill, the town, Oct. 14, appointed Ens. Parker and Timothy Robinson to treat with him. It is not charged that he took illegal toll, but the toll was thought exorbitant.*[1]

In 1767, the town demanded another mill and "voted to build a dam at William Green's River; if Benj. Gifford will build the mill and keep it in repair,—he to make the dam and a sufficient cartway." [2] As the town grew, it continued to sponsor mills. In 1788, permission was granted to Shubael Lawrence to set up a fulling mill at Dexter's River. His mill was subsidized by being exempt from taxes.

A year later, in 1787, Falmouth's most famous windmill was built by Jesse Gifford in West Falmouth for Samuel and Joseph Bowerman and Richard Lake, and remained in the Bowerman family until 1816. The Bowermans were Quakers and, as the *Boston Sunday Globe* later put it, Joseph was especially kind: "It will be at once imagined that Joseph Bowerman was benevolent, but few would surmise that he used to carry apples to the mill to give to the children who came to see him grind."[3] Theirs wasn't the only windmill in Falmouth, however; state records show that in 1800, there were eight mills in town—one fulling mill and seven gristmills—and that most of the latter were windmills.

In August of 1887, seemingly all of Falmouth turned out for "A Unique and Pleasing Centennial Celebration" of the old windmill. The *Boston Sunday Globe* reported, "They participated in the dangers of the funniest kinds of athletic sports and had music and literature almost under the shadows of the great sails." Phear Hamblin Baxter, "a good old lady who lives down here a short way," told stories about early millers Barnabas Hamblin and his son, Sylvanus.

Silas F. Swift, the miller at the time of the Centennial, noted:

> *The grist is uncertain. It will vary all the way from 50 to 75 bushels* {a day} *in a good breeze. I have ground as high as 100 bushels in a day. It don't pay to set*

the mill going for less than a grist of 10 bushels. The miller takes all of his share for his stock; I've got more'n enough stock to eat mine. The business now don't keep us busy more'n one day in the week on the average. I expect it's the oldest mill on the cape. The one at Falmouth and the one at Pocasset are both younger, but they're both hung up. There's one on Nantucket, I understand, that's older; but I doubt if it's put to much purpose except for exhibition.[4]

The morning of the celebration broke sultry, and the decorating committee arrived early. By 9:00 am, the four sails of the mill and its tail had been ornamented with U.S. flags and bunting. Chinese lanterns hung in windows and doorways, and the sails were festooned with garlands. A canvas banner on the side of the mill read:

Ye come and go, ye die and are born;
I stick to my work of grinding the corn.

The crowds assembled, with the men "in business-like jumpers and overalls as well as in knickerbockers, and women in neat gingham as well as in flannel yachting suits."[5] A long day of sporting events was planned, in fact so long that some matches such as tennis petered out and were halted due to the exhaustion of both players and spectators. There were potato races, egg races, wheelbarrow and three-legged races, an obstacle race, horseshoe pitching, and rifle shooting. Prizes included a potato masher, two three-legged stools ("a little too small to milk with"), and a photograph of Mrs. Grover Cleveland.

The night ended with fireworks, the release of three red balloons, and a dance accompanied by the Falmouth Band. One of the day's highlights had been a lengthy verse read by a Major Stevens. The Major's tale began, for no apparent reason, with the building of the Hillsboro Bridge in New Hampshire. Nearly a dozen verses later he made his way to the subject at hand. To the old West Yarmouth mill, he declaimed:

...Grind on, oh shrine of hungry pilgrim homes,
Grind on, all needless of thy aching bones,
Grind on, unmindful of the rhymester's chaff,
Stern in thy labor, tho' the worldlings laugh
Advantage take of every wind that blows,
In summer's heat, through winter's frost and snows.

...The poet's task already now is done,
While thine we hope is only but begun,
And when another hundred years shall pass,
When homes that know us now, no more alas
Shall wait our coming at the day's decline,
Joy in our joys, and at our fears repine;
Around thy walls shall stand another throng,

While we by time's swift current borne along
Life's journey's end have reached, may then look down,
I trust not up—and see that ancient crown
Of thine, oh, aged, quaint, industrious mill,
Casting the same old shadow on the hill.[6]

But Major Stevens was wrong. No one would gather a hundred years later to celebrate the mill's Bicentennial. In 1922, Seth Gifford bought the mill and very soon thereafter sold it to a Brockton, Massachusetts, ice cream store and restaurant known as Dutchland Farms. A windmill seemed to fit its image, so the mill was gutted and carted to Brockton. It later burned to the ground in a fire started by either a careless worker or a homeless person sleeping in the mill—the evidence disappeared with the windmill.

The Barnstable Village Windmill

Barnstable, Cape Cod, Massachusetts

One of the earliest documented windmills on Cape Cod was raised in Barnstable in 1687. Though the mill no longer stands, the contracts left behind are the best description we have of how Cape Cod windmills were built.

In the 1620s, the area around Barnstable Harbor, known as Cummaquid, was under the control of the sachem (tribal chief) Iyanough. In 1639, the Reverend John Lothrop and his congregation from Scituate set up housekeeping there. Iyanough's reaction went unrecorded, but a letter from Lothrop to Governor Thomas Prence asks him to "make composition with the Indians for the place with what speed you can, and we will freely give satisfaction to them." [7]

My favorite Barnstable story goes back to the summer of 1621 and involves the scandalous Billington family. The Billingtons made quite an entrance into the New World aboard the Mayflower in 1620. They had, in fact, nearly aborted the Mayflower's celebrated arrival. The two young Billington boys, John and Francis, had almost blown up the ship while it was anchored at the tip of the Cape. They had lit squibs and shot off muskets in their father's cabin, with an open keg of gunpowder only a few feet away.

Later in 1621, according to Governor William Bradford's first-hand account of Plimoth, seven-year-old John Billington "lost himselfe in ye woods & wandered up & down some 5 days, living on berries & what he could find." He had wandered some twenty miles out among the Cape Indians. Pilgrim Edward Winslow sailed to Barnstable Harbor with a search party and was taken ashore by Iyanough, who happened to know where the boy was. John Billington's reunion with the Pilgrims was quite a scene: The boy, decked out with beads and shells, was paddled out to the Pilgrims' boat by the natives. Iyanough saw the whole troop off with presents and dances.

Bradford had considered the Billingtons "one of ye profanest families amongst them." Young John never grew up, having died of gangrene in 1638. His father, John Sr., suffered a worse fate. He had been in trouble for dueling and was tied head-to-heels as punishment in 1623. In 1630, he was the first man to be hanged for murder. As Bradford summarized, "His fact was that he waylaid a yong man, one John Newcomin, about a former quarell and shote him with a gune, whereof he dyed."[8]

Barnstable's brush with the Billingtons had no lasting effect, and the town thrived. Stretching from Barnstable Harbor on the north to Nantucket Sound on the south, it is the most complex and varied of Cape Cod towns, being, in fact, comprised of seven separate villages: Barnstable, Centerville, Cotuit, Hyannis, Marstons Mills, Osterville, and West Barnstable.

Barnstable Village had at least two windmills, built in 1687 and 1785, though none exist today. At the town meeting of January 19, 1687, the people of Barnstable voted to build a village windmill. The amount of thirty-two British pounds was appropriated, as well as the land for it: "five acres of upland and as much marsh." The town chose the best millwright available, Thomas Paine, and drew up a detailed contract, a fascinating very early description of a Cape Cod windmill's construction. The Town Records of 1687 state that Thomas Paine:

> *...shall make erect build and set up one Good Substantial windmill In some convenient place in Barnstable afores between ye Dwelling House of Nathel Bacon on ye east side of ye hill commonly called Cobb's Hill and shall and will provide and find all sorts of timber suitable for ye building of sd Mill, boards shingles nails and Mill Stones and Cloathing all sorts of Iron work belonging thereto and ye said Mill to keep maintain for ye space and term of twenty years in so good a capacity as Shee may Grind all the Corn of the Inhabitants of ye Town of Barnstable well and and as It ought to be Ground that they shall have occasion to be Ground that they may not be put to straights upon that account and shall and will well and workmanlike fraim Erect and Set up sd windmill and finish ye same that so Shee may Grind ye Corn of ye Town between this and ye last day of Octor next coming after ye date hereof In consideration of which sd windmill to be built set up done and finished in all Respects as above sd In manner and form aforesd...and In full payment of ye sd Sum of two and thirty pounds when the sd Wind Mill is in every Respect finished and doth Grind Corn well and further that sd Mill shall be ye Proper Estate of ye sd Thomas Pain or his Assigns and ye Ground It shall stand upon...Room to set a house upon shall and will make a Legal Conveyance of four or five acres of Upland and three or Four Acres of Marsh of ye Town's Commons unto ye sd Thomas Pain or his Assigns and shall and will Draw or cause to be Drawn ye Millstones to ye place where sd Mill is to be set up and if it should happen sd Mill Stones cannot be procured nearer ye Town of Eastham than to draw sd Millstones from Satucket Mill to ye place where sd mill is to be erected.*[9]

This lengthy document goes on for many pages, exhibiting a firm grasp of legal language but gleefully abandoning, as was common, the strictures of sentence structure and spelling.

The Dennis Port Windmill

Dennis Port, Cape Cod, Massachusetts

Authentic windmills are gone from Dennis Port. Reproductions, like this one from the 1930s, have replaced them.

The five villages of Dennis are known for their colorful sea captains, like the lugubrious Joshua Sears. Or Joseph Baxter, who could have become the King of Saipan, but when offered the hand of the king's daughter said, "To become a king had never been any part of my ambition...." [10] Dennis windmills, less celebrated than Dennis sea captains, seemed to have unusually bad luck. Like many old windmills, some were dragged away to other towns. Others were doomed to more bitter destinies: pilfered for lumber, struck by lightning, or the Dennis Port windmill burned to the ground by young Fourth of July revelers.

In 1979, windmill historian Jim Owens asked an innocent question about an old postcard and got some rather surprising answers. Owens was presenting a lecture on "Cape Cod Windmills" at the Dennis Historical Society when he paused during his slide presentation to offer a mystery: "I'm stopping for a while at this point for a reason," he said. "I made this slide from one of those tinted postcards produced early in this century. I want you to study it for a moment. The building you see here is identified only as a Dennis Port windmill. I'm showing it in the hope that at least one of you can tell me what was its purpose and where it stood."

Several people in the audience stirred, then Dennis Port native Edward P. Chase rose and declared, "That was our bathhouse! Most Dennis Port boys used it to undress and dress in when we went swimming. We used it until some older fellows burned it down on the night before the Fourth. I remember it well. It stood on Chase Avenue not far from the foot of Depot Street."

Owens was speechless—boys using an old mill as a bathhouse was one thing. Arson was quite another.

"That building was the windmill that pumped water for the old saltworks where the Dennis Shore Cottages colony now stands," Chase elaborated. "But I wasn't kidding. When it was no longer in use we boys took it over, as I said, as a bathhouse. Not many houses around there then; nobody bothered us."

Lester M. Edwards corroborated Chase's story. Born about 1895, even earlier than Chase, Edwards added further details: "Played in and around that old mill a lot when I was a boy. The upstairs was a fine place to play, even if it was sort of spooky and cobwebby, but it was downstairs that we got into our bathing suits. That is when we didn't go in the other direction and change behind the barrels piled up back of one of the fish shanties over there at the end of Tom's Path. But the mill was best."

One of many Dennis windmills, it was believed by old-timer Nathaniel Wixon, town selectman and fisherman, to have been called the Reuben Burgess Mill. The records of miller Burgess are of little help in substantiating this. He died in 1885, but no windmill is listed in his estate. As to the mystery of how the windmill burned down, reporter Ellis Morris interviewed Lester Edwards for the *Yarmouth Register* newspaper and Edwards revealed, "There wasn't much in the way of fireworks in those days and certainly nothing of the sort was sponsored by the towns. Bonfires mostly. Big ones, too. Sometimes old barns and outhouses."

Pressed for details on who set the blaze, Edwards allowed, "Yup—I do know. But can't tell. Won't tell. Some of the those boys who did it must have relatives still around. Can't tell on those boys, you see."

The Hatsel Kelly Windmill

East Dennis, Cape Cod, Massachusetts

The story of the Judah Baker Windmill has already been told—of its construction in South Dennis in 1791 and its move to South Yarmouth in 1866. The story of the reverse journey of a mill is also worth telling, of a windmill built in Yarmouth in 1766 and moved to East Dennis in 1775. However, unlike the Baker mill, which still sits neatly by the Bass River, the East Dennis windmill was felled by lightning.

We're fortunate that a handwritten account of the mill remains in the records of a meeting of the Pilgrim Club, held in East Dennis on April 24th, 1896. On that night, Ray E. Mooar told of the first time he noticed the ruined mill tower:

> *In company with a friend I made the ascent of the hill and stood beside an old windmill long past its usefulness and fast going to decay. Originally octagonal in shape the sides had fallen away in several places and nothing stood in the way of*

entering unless it might be a fear lest the whole structure come tumbling down upon our heads. But upon further examination the supports were found to be still secure and we entered.

Directly above were the great mill stones and the trough which formerly carried the golden meal into the bins below. Here was the end of the shaft set so as to turn in an iron socket. On one side were the remains of a lever, connected above with a band to check the speed of the wheel and bring it to a standstill. All around were pieces of broken boards, shingles, nails, &c. Going up the stairs to the second story the stones were in full view. The boarding around these, however, was very insecure and rendered progress somewhat dangerous. The great beam to which the arms were attached was still in position.... The cog wheel on the shaft was gone....

From the windows a beautiful view of the bay could be obtained lying blue and clear in the sunlight. It made one think with sorrow, that this old historic building which had seen so much of the past could never again respond to the gales that still come from off the bosom of that restless water; that the whirr of those busy stones was silent forever; that the life of that old structure was in the past along with that of many who had lived beside it and known it well.

Determined to learn the facts of the mill, Mooar had sought out old William Sears, who had been born about 1808:

On going to the house, I was cordially welcomed. I found Mr. Sears to be a veritable repository of all the old facts connected with the history of the town...his eighty-eight years of life give him more of the past than many men now living can claim, and his memory is still unclouded.

From Mr. Sears I learned the following facts concerning this ancient landmark. It was built in 1766 in the town of Yarmouth by a man named Hatsel Kelly. This same Hatsel Kelly was a famous builder of windmills on the Cape. After remaining there for some years in active operation, it was in 1775 removed to Dennis and erected on its present location....

The men who were interested in this mill were William Crowell, Peter Sears, Edmund Sears and John Chapman. The last named, Mr. John Chapman, was the grandfather of the present owner, Henry R. Chapman. It is said that he died in 1815, and on the day of his burial was carried two miles to North Dennis on the shoulders of men. In those days a man was far more truly a bearer than at the present time....

Abraham, the second son of John Chapman, tended the mill for many years. As the owners were dependent on wind for their business, they would take advantage of it whenever it came, and many were the nights used in this way when trade was rushing, and orders far ahead.

> *Thus affairs went on till 1869 when during a thunder storm the lightning struck the iron cog wheel which had replaced the old wooden one on the shaft, and disabled the mill.... Repairs were attempted and for a time hope was expressed of putting it into running order again, but time wore on and the work was still unfinished.*
>
> *Years have passed and now nothing remains but the shell of what was once the pride of its possessors. The good oak timber will last years yet and is a substantial monument to its maker builder.*
>
> *All in all it is a most unique relic and well worth a visit from all who are at all interested in the mills of Old Cape Cod.*[11]

It has been well over 100 years since Ray Mooar wrote of the Hatsel Kelly windmill. It stood for generations off Route 6A behind Worden Hall, but no longer exists.

Of the many lost windmills of Dennis, one in particular must be mentioned, for it was visited by Henry David Thoreau. Thoreau's classic book, *Cape Cod*, is the best account of the daily life, characters, history, lore, and sensual beauty of the Cape. In October, 1849, Thoreau and his friend William Ellery Channing set out from Concord for an excursion on the Cape. While walking through Dennis, they visited Uncle Rufus Howes' windmill. Known also as the South Mill, it had been built in the late 1700s. It stood west of the Burying Ground on what is now Route 6A, and was taken down in 1874 by its last millers, Rufus and Edmund Howes.

Thoreau doesn't write of his impression of the Howes windmill in either his journal or in *Cape Cod*. But, as recorded in Nancy Thacher Reid's *Dennis, Cape Cod: From Firstcomers to Newcomers*, Thoreau left a vivid image behind. Isaac Freeman Hall (1847-1928), remembered being told about Thoreau's visit and of the writer's comments. According to Hall, Thoreau said he "found great beauty in the flour-encrusted cobwebs which hung from the windmill's rafters." [12]

The Truro Highlands Windmill

Truro, Cape Cod, Massachusetts

I drove to the Truro Congregational Cemetery looking only for the Snow Family grave site, with its old millstone as a family marker. I found it at the peaceful top of a slight hill. The stone, from a Truro tidal mill, marks the graves of Charles W. Snow (1855-1936), Abbie H. Snow (1858-1919), and Anna G. Snow (1885-1891).

But just down from the Snow Family plot, I spotted another millstone by a grove of locust trees. This discovery led me to the stories of Truro's most eminent windmill, the Highland House hotel, and the town's most prominent musical duo.

Carved near the upper rim of the millstone, I found the acronym, "JOBI." Below it, a plaque reads, "Joseph A. 'Joe' Colliano (1926-   )" and "Wills C. 'Bill' Hastings (1920-   )." Below their names is the following inscription: "...and after this brief

The Truro Highlands Windmill, from *Ballou's Pictorial Drawing Room Companion*, 19th century.

intermission, the music will continue." Small footstones, flush with the ground, are decorated with g-clefs. I wondered what "JOBI" referred to and what kind of musicians Colliano and Hastings were. Most of all, I wanted to know which mill once housed the honey-colored millstone, and the part that mill might have played in their lives. A phone call to Joe Colliano and Bill Hastings solved the mysteries. While caring for Bill, who was in failing health, Joe took the time to tell me of their fifty-one years in Truro.

The Snow Family grave site with millstone marker, Truro Congregational Cemetery, MA. Photo by the author.

Joe, a pianist, and Bill, an organist, had met in New York's popular music scene. In 1951, they were living in Troy, New York, and performing in the Albany-Troy area. That summer, they went to stay at the Highland House hotel in Truro. After the arrival of the railroad in the late 19th century, the Truro bluffs were dotted with small hotels. Located by the famous Highland Lighthouse, the Highland House, now the home of the Truro Historical Society, is the last one left standing on the Truro Highlands. As Joe recalled,

A millstone from the **Truro Highlands Windmill** marks the grave site of Joseph Colliano and Wills Hastings, Truro Congregational Cemetery, MA. Photo by the author.

a room in 1951 went for $3 a night. He and Bill were so taken by the hotel and the location that they offered to play organ and piano there.

Eventually, Colliano and Hastings were invited to take over a combination souvenir and hot dog stand nearby the Highland Lighthouse. Joe and Bill began selling their own pottery, and it became so popular and collectible that the shop evolved into Jobi Pottery (from the first two letters of each man's name). As if that weren't enough, Joe and Bill earned their permanent place in Truro history by becoming the final managers of the Highland House hotel from 1964 to 1969. While there, they noted the millstones left at the side of the hotel and learned of an old windmill that had long vanished from the site.

The windmill on the bluffs had long been forgotten in the shadow of the Highland Lighthouse. In 1797, the U.S. government bought 10 acres of the Truro highlands and built the first of Cape Cod's lighthouses. Highland Light was the first beacon visible to ships making their way to Boston Harbor from Europe, and today is one of the most famous of Cape Cod landmarks. The uncelebrated windmill was an octagonal smock mill, with boat-shaped cap. Extending from the cap was a wooden windshaft on one side and, on the other, a tail pole with a cartwheel on its lower end. A fish weathervane topped off the cap.

Truro had a large number of the small windmills used for saltworks, and three or four traditional grain windmills. The Highland Windmill has been colorfully described by Shebnah Rich in his *Truro – Cape Cod* (1883). Rich also offers a rare portrait of one of the last of Cape Cod's windmillers:

> *It was built early in the present* {19th} *century. The frame was Southern tun timber, with its huge dimensions, as landed from a vessel stranded on the back of the Cape, and drawn to the hill-top by ox teams. A wash-barrel of grog was used at the raising. Doubtless this frame would have stood a thousand years, as when, two years since, removed it was as sound and bright as when first lifted to its airy summit. Its tall arms, and one long leg, the mast of a dismantled schooner, which, like a huge spider's web, angled from the cornice to a little wheel on the ground, nearly a hundred feet distant, and which turned the mill's arms windward, were the outside wonders.*
>
> *Inside, the giant shaft, the remorseless cogs, the iron spindle, the upper and nether millstones, were wonders of mechanism, and fill my mind with admiration for the men who could construct such mighty engines of power and cunning. On the wall of the first loft, nailed against the timbers, was plain deal board about eighteen by forty inches, on which, carved by some educated pen-knife, were these letters,—F.A., A.H., S.R. Which being interpreted meant, "This mill is owned in equal parts by Freeman Atkins, Allen Hinckley, Samuel Rider." These were important personages in my mind. The first of them only concerns us in this sketch.*
>
> *He* {Freeman Atkins} *it was who climbed the slender latticed arms and set the sails; he it was who hitched the oxen, waiting grist, to the little wheel, and with the boys pushing, turned the white wings to the wind's eye; he it was who touched the magical spring, and presto! The long wings beat the air, the great shaft began to turn, cog played to its fellow cog, and the mammoth stones began to revolve. He it was who mounted like Jove upon his Olympian seat, and with one hand on the little regulator, that, better than the mills of the gods that ground only slow, could grind fast or slow, coarse or fine, with the other hand caught the first golden meal.*
>
> *I see him now, in my mind's eye. A tall man, with long arms, like his mill; kindly blue eyes, angular face, prominent nose, and close shut lips, with a lingering of the old quarter-deck compression still revealed.... Ever and anon as he removed his great bony hand from the hot meal, touching his face and long nose till whitened like a distant promontory, he grew still more attractive, and I nestled still nearer his coveted seat, encouraged by his kindly manner to ask a few more questions.*
>
> *This old man with his grist mill, and salt works, and patch of cornfield, and semi-weekly newspaper, was no every-day man. In his younger days as sailor and master of a ship, he had seen the world....*
>
> *The old miller has ceased from his labors, and the sound of the grinding is low. A stone to his memory in the Congregational churchyard has the following:*

"He came down to the grave like a shock of corn fully ripe.
In memory of
Capt.
Freeman Atkins
died
Nov. 1, 1855,
Aet 79 years 1 mo
& 14 days.
Though time had set the seal of years upon his brow,
Yet still that brow bowed not beneath the weight of care,
Till by the reaper's hand in Death's embrace laid low,
Like shock of golden grain well prepared."

I asked Joe Colliano about the saying on his and Bill Hastings' headstone in the same cemetery. He told me that when he and Bill were young they used to listen to the Big Bands playing on the radio. Periodically, the music would pause and the announcer would say, "...after this brief intermission, the music will continue." Joe said that both he and Bill believe we've all been here before and we will be back. Within weeks of speaking to Joe, in the early summer of 2002, I learned that Bill Hastings had passed away.

Woodcut of a Truro windmill, from Shebnah Rich's *Truro–Cape Cod.* Truro, Massachusetts, 1883.

Provincetown's Lost Salt Making Windmills

Provincetown, Cape Cod, Massachusetts

One day in early June, I set off by kayak from Provincetown Harbor in search of the lost village of Long Point. If Cape Cod is an arm, Orleans its elbow, and Provincetown its curled fist, Long Point is the thin nail off the last finger of Provincetown. It's the spot that most brings to mind Thoreau's conviction that on the Cape, "A man may stand there and put all America behind him." Few realize that at this dramatic point of land once stood six windmills and a precarious village.

There are only two ways out to Long Point today. On foot, one can hike across the breakwater at the far west end of Commercial Street, making sure that the tide is low,

A 19th century woodcut of Provincetown, MA, with salt making windmills.

or one can go by boat. Keeping Long Point Lighthouse straight ahead, I crossed the harbor in my yellow kayak in the wake of fishing boats and whale watching vessels. The water turned turquoise as I approached the sandy spit of Long Point and beached my boat.

Today, all that is left of the former settlement is the little lighthouse, built in 1826, and two mounds of earth that used to support Civil War forts. (When the Confederate Army proved uninterested in attacking Provincetown, the forts were dubbed "Fort Useless" and "Fort Ridiculous.") Otherwise, the isolation is sublime. I walked this farthest reach of the Cape and imagined what it would have been like to be one of the thirty-eight families who formed the village here in the 1800s.

The first house was built on Long Point in 1818 by John Atwood, who was attracted by the abundance of cod, mackerel, shad, and bass. Other families followed and soon there were more than twenty boats engaged in cod fishing alone. Atwood built a wharf on the point's north side for the Cape Cod Oil Works. Miss Hannah Sanborn kept the first school on Long Point in the lighthouse. It began with only three students, but by 1846 there were sixty pupils and a separate schoolhouse was built. The schoolhouse also served as the settlement's church on Sundays, when a minister occasionally rowed out to preach. As Josef Berger said in his *Cape Cod Pilot*, "It was an exciting neighborhood to live in. Children who might have been afraid of dogs elsewhere, here ran from the sharks."

Long Point was an ideal spot for windmills, and soon the lucrative business of making salt flourished there. The population peaked at over two hundred people, many of whom tended the saltworks. Eldridge Nickerson built the first windmill and saltworks. At one time there were six windmills pumping water into seven or eight thousand feet of evaporation vats. The shores of Provincetown were, in fact, lined with windmills. In 1837, there were 78 saltworks throughout the town, producing more salt than all other Cape Cod towns save Dennis and Yarmouth.

Print of Provincetown, 1836.

By 1850, "the question whether they were living on land or sea grew so perplexing," as Berger wrote, that some of the families began moving off Long Point. When the Civil War broke out, only two houses and the schoolhouse remained. The homes were moved off the point on scows and set up on Commercial Street; small plaques presently identify each of them. The schoolhouse, too, moved to Commercial Street and was later used as the town's post office. A story is told of the shipwreck of the *Nina* in 1871. A tidal wave drove the ship ashore, clear through the post office wall. The skipper jumped out and mailed some letters.

The most poetic description of saltworks and their windmills was written by Henry Thoreau for his book, *Cape Cod*, while on his 1855 trip to Provincetown:

> *This was the very day one would have chosen to sit upon a hill overlooking the sea and land, and muse there. The mackerel fleet was rapidly taking its departure, one schooner after another...like fowls leaving their roosts in the morning.... The turtle-like sheds of the salt-works were crowded into every nook in the hills, immediately behind the town, and their now idle wind-mills lined the shore. It was worth the while to see by what coarse and simple chemistry this almost necessary of life is obtained, with the sun for a journeyman, and a single apprentice* {the windmill} *to do the chores for a large establishment. It is a sort of tropical labor, pursued too in the sunniest season; more interesting than gold or diamond-washing, which, I fancy, it somewhat resembles at a distance....*
>
> *It is said, that owing to the reflection of the sun from the sand-hills, and there being absolutely no fresh water emptying into the harbor, the same number of superficial feet yields more salt here than in any other part of the country. A little rain is considered necessary to clear the air, and make salt fast and good, for as paint does not dry, so water does not evaporate, in dog-day weather. But they were now, as elsewhere on the Cape, breaking up their salt-works and selling them for lumber.*[13]

Martha's Vineyard's Lost Windmills

Martha's Vineyard, Massachusetts

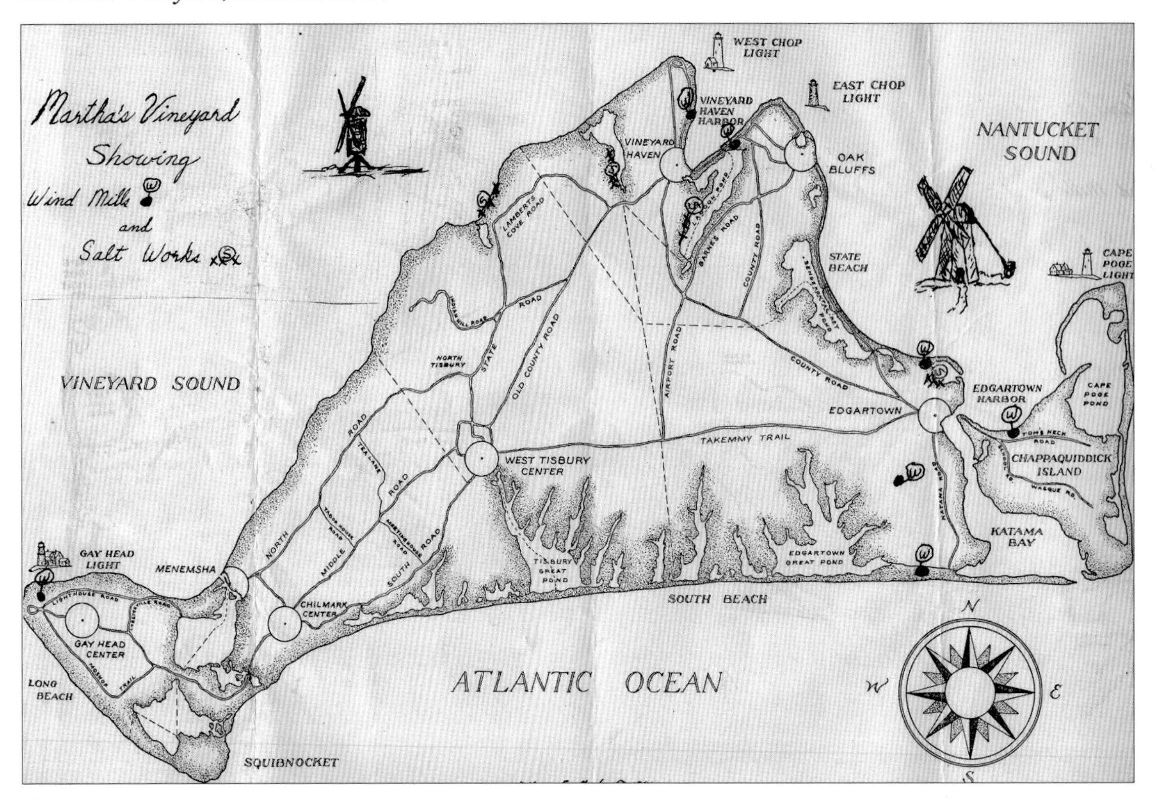

Map of Martha's Vineyard with windmills and saltworks. Copyright William Marks, from his book, *The History of Wind-Power on Martha's Vineyard*, 1981.

Subject to winds from all sides, Martha's Vineyard is one of the finest wind-gathering points in the country. The island, and in fact all of Cape Cod, has a wind speed of 14 to 16 miles per hour, higher on average than most of the United States. From the islands's earliest days of settlement, the Vineyard's windmills pumped sea water for the making of salt, and ground corn, chocolate, snuff, mustard, and bark for tanning leather.

In 1807, when James Freeman published his book, *Description of Dukes County*, there was a thriving salt making industry, powered by windmills. Freeman wrote, "Besides these mills there are, for the grinding of corn, four windmills in Edgartown, one of them on Chappaquidick; one windmill and three watermills in Tisbury; and five watermills in Chilmark."

Historian and environmentalist William Marks has turned up much evidence of the island's saltworks and their integral windmills. An 1807 survey counted salt, corn, and wool as the three most important Vineyard industries, all of which required mills—water mills for wool, and windmills for corn and salt. Marks has found deeds for saltworks located on the beach at Lambert's Cove, and a diary entry of Jeremiah Pease for December 10, 1848, noting the removal of the vanes on a saltworks windmill. The latter is evidence of the seasonality of windmill-driven saltworks. Marks notes the importance of the salt industry to Martha's Vineyard, and how the saltworks

An up-island farmstead and windmill, Martha's Vineyard, mid-20th century. Photo courtesy of William E. Marks.

were devastated during the Revolutionary War. The British destroyed the saltworks and captured salt supplies at Edgartown during Gray's Raid of October, 1778. Other British raids, like that of George Leonard, Esq., in 1779, burned ships and saltworks in several Vineyard locations.

An early 20th century photo of Edgartown Harbor, Martha's Vineyard, with windmill water pump on right. Photo courtesy of William E. Marks.

In an essay written for the Duke's County Historical Society in 1956, historian Eleanor Mayhew noted several windmills in Edgartown (the "Town Mill" between the 18th century houses of John Coffin and Joseph Kelly), and Gay Head. Of Oak Bluffs, she wrote, "A burying ground for sailors may be found here, but most of the markers, which were made of teak brought home on a whaler, have disappeared. There were early brick kilns in this vicinity, and a windmill." [14]

A windmill and farmstead in West Tisbury, Martha's Vineyard, mid-20th century. Photo courtesy of William E. Marks.

The Harbor Windmill & The Littlefield Windmill

Block Island, Rhode Island

The Nature Conservancy calls Block Island "one of the last great places in the Western Hemisphere." Seven miles off the Rhode Island coast, it's an idyllic glacial wonder, formed 12,000 years ago. With 365 freshwater ponds, 32 miles of trails, and 17 miles of beaches, it's well worth a visit—but not for its old windmills.

One of Block Island's lost windmills.

Block Island's two most historic windmills are gone. One left a sad story of lost opportunity, the other a rich heritage in the writings of its last miller. Presently, the subject of windmills on the island is one of controversy and lawsuits, as residents balance the desire for renewable wind-powered energy against property values and landscape preservation.

In 1891, the Harbor Windmill and the Littlefield Windmill still stood, but just barely. The local paper published this lament:

> *The two windmills, one on the road to the Center and the other on the West side, form a unique and striking figure in the landscape. Not that there is anything peculiar or uncommon in the use of wind as a motor, but simply that the odd, fantastic appearance of the mills themselves, their heavy oak frames, the moss covered shingles, and the very air with which, with their long vanes that nearly sweep the ground, they mount guard over the surrounding country, seems to imply a consciousness of superiority over the neighboring edifices, based upon an old age following a life of unchallenged usefulness.*
>
> *Generations have come and gone, children have grown up to manhood and descended the hill of life, some tarrying until perchance, their own grandchildren were gray-headed men, and still the tireless vanes go round.*
>
> *But Father Time, who also goes his rounds with the same tireless persistence, and whose sickle, first or last, falls on all created things, animate or inanimate, has for some time had his eye upon the old mill. Verily, they have nearly fallen into a state of "innocuous desuetude."*[15]

The Harbor Mill had been built off-island in the 1770s and brought to Block Island around 1810. The 35-foot-tall smock mill, with domed cap, ground corn until 1891. Ten years later, a preservation group formed and bought the mill and two surrounding acres. In 1903, the local newspaper reported:

> *The top, which had blown off, had been preserved and is ready to be put in place again. It is the intent to put the mill in perfect repair, and have it grind corn again...to make it an educational example of what was done in more primitive times. The mill is to be used as a sort of museum, where curios of all kinds, Indian relics, interesting souvenirs of the boats which have been wrecked on the Island, relics of old sea captains, etc., will be kept.*
>
> *It is part of the plan to transform the land into the most attractive park, where trees, rare flowers, lotuses and aquatic plants shall grow luxuriously.... The committee is now at a standstill for lack of funds, and it has decided to hold a fair....*[16]

Funding couldn't be found, however, and what was left of the abandoned windmill came crashing down during the Hurricane of 1938.

The Littlefield Windmill is believed to have been built in Fall River, Massachusetts. Mill historian Robert M. Downie writes that by 1815 the mill was located on Center Road opposite the intersection with Beach Avenue. When funerals passed by the windmill on the way to the island cemetery, the blades were stopped as a token of respect to a departed islander.

In 1877, the Rev. Samuel Livermore of the Block Island First Baptist Church wrote:

> *This mill, owned by Hon. Ray S. Littlefield, is capable of grinding one hundred bushels of excellent cornmeal in a day when the wind is favorable.... Many* {summer visitors} *will remember with pleasure the Littlefield Mill, so near the Central House, and in and around which the children have played in summer, and within whose dusty walls some of them have been gathered for an hour's Sabbath-school, where they have sung their familiar hymns and recited their lessons....*[17]

In 1888, John Edward Littlefield and Gilbert Sprague moved the mill to Old Mill Road on the West Side. John Littlefield, Block Island's last windmiller, had been writing a column for *Cooley's Weekly* since 1875. In his column he left us the best description of what it was like to run a windmill at night—a not uncommon task when every bit of the fickle wind was used, whenever it blew:

> *Another arm has been placed on the mill. Last Saturday we picked the mill stones, commencing early, and at 2 p.m., having finished and swept up, and a fine breeze was blowing from the southwest, we started up intending to "grind out." We had had no wind for some time and of course could not grind any of the corn which had been brought in. Everything went well and when darkness fell we lighted up and prepared to spend the night on the mill hill.*
>
> *Around nine o'clock heavy clouds hung over the western skyline, split occasionally by lightning flashes, and an hour later the wind hauled around into the west with black clouds hovering over Watch Hill, gradually working to the north and east accompanied by much vivid lightning. Now was the time to keep a good lookout. We were working under full sail and the wind was still very light. Around eleven p.m. the lookout reported "squalls" nearby and all hands turned out to haul from west to north and then to east.*
>
> *A single reef was taken all around and the lights, which the wind had extinguished, were relit again and in the murky darkness we resumed grinding. However, the wind gradually died down and at half after midnight we were in a dead calm with still four bags of corn to grind.*[18]

In the 1880s, steam power arrived on Block Island, and wind-powered windmills were in decline. In February, 1888, the *Providence Journal* reported, "Considerable difficulty is experienced by people here in getting table meal ground from the Island corn. Both the old windmills are badly out of repair and practically worthless, and it does not pay Mr. John F. Hayes, who put in steam power for the purpose of grinding corn, sawing lumber, etc., enough to make it an object for him to grind, so the citizens are in a dilemma." [19]

The Littlefield Windmill continued to grind throughout the 1890s until it was finally abandoned after 1900. All that remains are its millstones, one on either side of the road leading to the airport; and the foundation, looking like a pagan stone circle, where the old windmill once stood.

Harkness Windmill

Honyman's Hill, Middletown, Rhode Island

Harkness Windmill, with the miller and his family. Middletown, RI, 1900. Photo courtesy of the Middletown Historical Society.

The *Newport Mercury* reported on April 16, 1842:

> *An explosion occurred last Saturday week at Mr. Thomas Harkness' windmill near the beach, which besides other damage came very near proving fatal to Mr. H. The mill was under full headway at the time, and but a few minutes previous to the explosion Mr. H. replenished his hopper and went downstairs. Soon after, not more than three minutes, he heard a loud crashing noise above and proceeded immediately with due caution outside to stop his mill, in the usual way, when he discovered that the millstone had exploded into parts and burst through the sides of the mill, a very substantial building, and was laying partly embedded in the ground, one half about thirty feet, the other half about ten feet distant from the mill. The weight of the stone we understand was something like two tons.*

The Harkness Mill burned to the ground in November, 1905.

The Old Saybrook Windmill

Old Saybrook, Connecticut

The most famous of Connecticut windmills was built in 1636 in Old Saybrook. Wishing to keep the Dutch off the coast, the English granted a patent to Lord Say and Sele, and Lord Brooke, for the land "lying west from the Narragansett River 120 miles on the Sea Coast and then in latitude and breadth aforesaid to the South Sea." In 1635, John Winthrop, Jr. (son of the governor of the Massachusetts Bay Colony), and his men reached Saybrook. They were followed by an engineer named Lion Gardiner, who was in charge of building a fort and a windmill at Saybrook Point. Today, there is a statue of Gardiner in his cuirass and helmet at the site of the old fort.

All that remains of the Old Saybrook windmill are two millstones which, if one looks carefully, can be found about a mile from Saybrook Point in the front yard of the home at 430 Main Street. This, apparently, was where the outer palisade of the settlement stood. A plaque there reads, "These stones are the last relic of the windmill which was built by Lion Gardiner on this site in 1636/7 and stood here for 175 years...."

The millstones of the 17th century **Old Saybrook Windmill**, CT. Photo by the author.

North of Old Saybrook along the Connecticut River is Essex, where I found evidence of one of the most unusual uses of a windmill. The Essex area became famous for goods made of ivory; in fact, a nearby town was named Ivoryton. The ivory works of G.W. Dickinson & Co., for example, made piano keys and a variety of ivory stationers' goods and notions.

Before a consistent supply of ivory became available around 1830, both horn and ivory were used. In 1798, Phineas and Abel Pratt invented a machine for making combs out of horn. According to *The History of Middlesex County*, "The manufacture of combs in this country was first begun by Phineas Pratt and his son Abel.... They were the first inventors of machinery for cutting the teeth upon combs, by which they could be produced so as to compete with English manufacturers. The shop in which they worked stood a few yards west of the site of Pratt's blacksmith shop, and the first machinery was driven by wind power."

Essex historian Donald Malcarne tells me the Pratt windmill was near Exit 3 off Route 9, and that it was torn down in the 1830s. Malcarne also notes the Hayden-Cheney windmill, built before 1780 and remodeled with metal blades more than 100 years later. The land where it stood is now the Cross Lots Nature Sanctuary on West Avenue.

Hatch's Mill

Windmill Hill Lane, Castine, Maine

Postcard of the historic marker for **Hatch's Mill,** Castine, ME. Courtesy of the Castine Historical Society.

Searching the Internet for windmills one day, I entered variations of "Maine" and "windmill." The only intriguing hit was a realtor's advertisement for a home being sold on Windmill Hill Lane in Castine. I contacted Castine Realty on the odd chance that someone there might know if a windmill had ever sat on the hill. Wanda Wood wrote back saying, "Yes, there was a windmill on this road, in fact, at one time there were three in our small town."

Castine is rich in history, having been possessed and fought over by the Dutch, the French, and the British. The resulting burning and pillaging during such struggles destroyed a lot of the oldest buildings. In many cases the only evidence of these structures and events are the historic signs marking their location. Maine writer Louise Dickinson Rich complained of one such sign as "the most annoying plaque I have ever encountered. It says that on this spot the Second Miracle in New France took place; simply that and nothing more. You are left all up in the air and consumed with curiosity. What *was* the Second Miracle? What and where was the First?"

Sally Foote, the archivist at the Castine Historical Society, confirmed for me that there was indeed a windmill on Windmill Hill Lane. Called Hatch's Mill, it ground corn in around 1800. The only other known windmills were put up much later—two by the water company to pump water, and one windmill garden folly built in the 1920s. Ms. Foote corroborated mill historians John White and James Owens, who described Maine as a place of many tidal mills and water mills, but few windmills.

A 19th century print of a windmill pump at Port Clyde, ME. From Samuel Drake's *The Pine Tree Coast*, Boston, 1891.

The Windmill Point Mill

Lake Champlain, Alburg, Vermont

Much like Maine, the green, wet, mountainous state of Vermont relied mainly on water mills. Still, taking a closer look along the shores of Lake Champlain, one finds Windmill Point in the town of Alburg. Contrary to expectations, there sits—and has

The Farris Windmill, West Yarmouth, MA, now in Dearborn, Michigan

Benjamin Godfrey Windmill, Chatham, MA

The Eastham Windmill, Cape Cod, MA.

Oyster Harbors Windmill, Osterville, MA.

The mysterious Newport Tower, Touro Park, Newport, RI.

First day issue of windmill postage stamps, February 7, 1980.

Some Surviving New England Windmills

Boyd's Windmill, Middletown, RI. Photo by the author.

Eastham Windmill, MA. Photo by the author.

Sur Mer Windmill, Chatham, MA. Photo by the author.

Old Higgins Farm Windmill, West Brewster, MA.
Photo by the author.

The Old Mill, Nantucket, MA, from the front.
Photo by the author.

New England Wind Farms

The Princeton, MA, wind farm on Wachusett Mountain. Photo by the author.

The Hull, MA, wind turbine at sunset. Photo courtesy of the Hull Municipal Light Department.

The Searsburg, VT, Wind Facility at dusk. Photo courtesy of Green Mountain Power.

It may not be too long before sights such as this countryside scene in Fenner, NY become commonplace in New England. Photo courtesy of GE Wind Energy.

The proposed wind farm for Nantucket Sound, MA certainly wouldn't be the first offshore wind farm. The facility pictured here is located in Utgren, Germany. Photo courtesy of GE Wind Energy.

sat since the mid-1800s—a lighthouse. However, in the summer of 1749, Swedish scholar Peter Kalm had traveled along Lake Champlain on his way from New York to Canada. In his journal he wrote of the Alburg settlement: "A windmill built of stone stands on the east side of the lake on a projecting piece of ground. Some Frenchmen have lived near it; but they left it when the war broke out, and are not yet come back to it.... The English have burnt the houses here several times, but the mill remained unhurt." [20]

In the same year, Captain Phineas Stearns noted the lonely scene in his journal: "...at the emptying of the lake into Shamblee {Richelieu} River there is a windmill, built of stone; it stands on the east side of the water, and several houses on both sides built before the war, but one inhabited at present." [21]

Around 1731, a French settlement was started at what is now the town of Alburg. The first building of any permanence erected was probably the windmill, made of stone and costing the hefty sum of $800. Some seventy years later, Windmill Point was witness to the flourishing, dangerous smuggling trade that operated along the border of the United States and Canada. Cattle, lumber, and potash were smuggled into Canada in ships like *The Black Snake*, which was unpainted and smeared black with tar to evade the U.S. revenue officers. This ship and others like it, in turn, smuggled British goods into the States. General Levi House held a detachment of the First Regiment of his Franklin County Brigade at Windmill Point in readiness to intercept such smugglers.

Windmill Point never entirely settled down, and it remained a risky promontory for passing ships. With the mill long gone, a lighthouse was put on Windmill Point in 1858. ●

Chapter Five

Curious Reproduction Windmills and the People Who Love Them

"The old house itself had not been large enough to contain their fantasies, so throughout the woods you might come upon a Japanese bridge, an overgrown Shakespearean garden, or a ruined fountain. But most wonderful of all was Great-Grandfather Horace's windmill."

Hila Feil, *The Windmill Summer*

Boothden Windmill

Indian Avenue, Middletown, Rhode Island

The remains of **Boothden Windmill,** owned by an assassin's brother, Middletown, RI. Photo by the author.

When Edwin Booth heard what his brother, John Wilkes Booth, had done he said, "It was just as if I was struck on the forehead with a hammer." At the time, Edwin, one of America's great 19th century actors, was in his last week at the Boston Theater. It was the morning after April 14, 1865, when Edwin's dresser rushed into his bedroom with the morning papers and shouted that President Lincoln had been shot, and that John had done it.

Toward the end of Edwin Booth's life, he looked back on the miseries of the Booth family—the assassination, bouts of drunkenness and madness, suicide and murder —and cried, "What a curse is on the Booths, and for what? We must go back beyond any records that I have found for the cause of all the horrors that are heaped on us!"

Later, Edwin Booth would build a retreat where he and his daughter, Edwina, could find solace on the Rhode Island coast. He built a windmill at the edge of the sea where he could gaze out in the afternoon, and where Edwina waited for him while he was out sailing.

Before the Booth family name became synonymous with Lincoln's assassination, they were known as America's premiere acting family. Father Junius Brutus Booth was the greatest tragic actor of his day. In his prime, on the strength of his name alone, Junius could fill any theater in America. But his life was plagued by drunkenness and fits of madness. In his later years, before his death in 1852, it was only with the help of his son, Edwin, that he was able to mount the boards at all.

As Edwin toured with his father, he himself scored great success in some of the same roles his father had played. His notorious brother, John Wilkes, joined the theater as well and for a time the Booth brothers (including another son, Junius II) vied to carry on the great family name in the theater. In 1864, Edwin's place in American theater was assured by his famous 100-night run of *Hamlet* in New York City. The next year, his brother John assassinated Abraham Lincoln in Ford's Theater in Washington, D.C.

Edwin retired briefly from the stage, and later went on to build Booth's Theater in New York, and to found the Player's Club. But tragedy continued to follow the Booth family. His wife died. His brother Junius II died. Of the latter's daughter, Marion, Edwin wrote, "She's a jolly dunce & should be incarcerated in some mild asylum for dunderheads." Later, Marion's brother, Junius Booth III, killed his wife and himself.

Edwin Booth's theater failed in April, 1883, and a month later Booth bought land for his retreat in Rhode Island. He had a "cottage" built in Middletown, about three miles outside of Newport. It is a large, rambling, two-story gabled house on a slope, at the bottom of which are the remains of a windmill. He named the place "Boothden." In a biography of Edwin Booth, *Prince of Players*, Eleanor Ruggles described Boothden, its windmill, and the life Booth and Edwina lived there:

> *In his yacht, christened the "Edwina," Booth cruised on the river and into the ocean as far as West Island or even beyond.... Near the river's edge stood the reproduction of an old Norman mill built from a sketch he had brought back from Europe. It made a landmark seen for miles, and on foggy nights when he*

was out in the yacht, Edwina climbed the spiral staircase that ran around the outside of the mill and hung a lantern in an upper window. A sailcloth hammock with perpendicular ends and a flat bottom was set up in the wide porch on the river-side of the cottage.... Booth would lie in it for hours, smoking and dozing.

Booth used half of the Boothden estate for farming. The windmill, in fact, was originally built as a hen house. When Booth was in residence at his Boston home, the overseer of Boothden would send him eggs from the windmill. This windmill/lookout tower/hen house had one other curious use: After Booth's death, an Episcopal bishop moved into Boothden and used the windmill as a chapel. The top of the windmill was destroyed by fire in the early 20th century, and today it remains as a charming, circular, one-story summer house.

When Edwin Booth's life ended on June 7, 1893, the fates provided a highly dramatic climax to the tale of the notorious Booth family. At the very moment when Edwin Booth's body was being carried out of The Little Church Around the Corner in New York City, Ford's Theater—where John Wilkes Booth shot Lincoln—collapsed, killing more than twenty people.

Joe Jefferson's Windmill

Aptucxet Trading Post Museum, 24 Aptucxet Road, Bourne, Cape Cod, Massachusetts

Joe Jefferson's Windmill, Bourne, MA, with caretaker Tom Gerhardt and daughter Natalina. Photo by the author.

Edwin Booth had been great friends with another actor, Joe Jefferson. In 1891, Booth spent the Fourth of July at Jefferson's summer estate at Buttermilk Bay, Massachusetts. Among the guests was President Grover Cleveland. It turns out that Jefferson—possibly following Booth's lead—had a windmill at his summer home, as well.

Today, Jefferson's windmill is featured at the Aptucxet Trading Post on Cape Cod. It may seem odd that there is a Dutch-style windmill at the Trading Post. Historically, no windmill existed there, and further, its style is unknown on the Cape. Until examined by experts in the 1980s, it was thought that Joseph Jefferson's windmill had ground corn in Holland and had been brought over to the States from there in 1870. But the story of Joe Jefferson's windmill has closer ties to Grover Cleveland and Rip Van Winkle than to cornmeal or the Dutch.

Joe Jefferson was a member of four generations of Jefferson actors. He was, in his day, a superstar. Born in Philadelphia in 1829, Jefferson was a genius as a comedic actor, celebrated especially for his portrayal of Rip Van Winkle. In a seventy-one year career he amassed a fortune and a circle of friends that included the great actors and politicians of the day. Toward the end of his career, he bought a large piece of land on Buttermilk Bay (on the north side of the Cape Cod Canal, within Buzzards Bay), and in 1889 built the Jefferson home. He gave each of his five sons, his daughter, and his sister sites for homes surrounding his and invited them to spend their summers near him.

He called the estate "The Crow's Nest" and it was a family retreat, much like the later Kennedy Compound. President Grover Cleveland had his home, Gray Gables, nearby at Buzzards Bay and became fast friends with Jefferson. The two were often fishing companions, and Cleveland invited Joe to his home to await the returns of his 1892 nomination to a second presidential term. Other frequent guests at The Crow's Nest included the Barrymores—Ethel, John, and Lionel. They participated in games, pantomimes, and Jefferson's famous July 4th fireworks parties.

On April 1st, 1889, a terrifying explosion erupted in the basement of the main house. Apparently, a tank of kerosene overflowed and ignited while it was being cleaned up. A surviving telegram that Jefferson sent to his family states, "My house at Buzzard's Bay burned to the ground. No lives lost, thank God." Shortly after sending it, Jefferson learned that someone had indeed died—his beloved cook, Ellen, who had been in his family for twenty-five years.

Jefferson's painting collection, including works by Sir Joshua Reynolds and Jean-Baptiste Camille Corot (as well as a palette used by Corot), was lost in the blaze. Undaunted, Jefferson rebuilt the house in 1894 and summered there until his death in 1905.

Records of the date Joe Jefferson had his windmill built, and who designed it for him, were probably lost in the 1889 fire. We know that Jefferson, an accomplished landscape painter, had his studio on the second floor of the mill. The property was later acquired by the 3M Company, and on January 27, 1969, 3M donated the windmill to the Bourne Historical Society.

The Society planned to use the mill as a gift shop at their Aptucxet Trading Post. In 1627, the Pilgrims at Plimoth dissolved their communal system and went into business in a big way to pay off their debts. They built boats and established trading posts, including the Aptucxet Post. They chose its location to conveniently trade with the Dutch at New Amsterdam (New York), and the Native Americans to the north and on the Cape. The Pilgrims traded furs, knives, corn, and tobacco; the Dutch sent sugar and Holland linen.

All that is left of the Aptucxet Trading Post is the foundation of the building, upon which the Bourne Historical Society has built a fine reproduction. The nearby Dutch-style windmill seems to serve as a reminder that this is one of the locations where the English and the Dutch intersected in the New World.

The moving of the windmill from Jefferson's estate entailed several problems—and one major blunder. First, between Buttermilk Bay and Bourne lie Buzzards Bay and the Cape Cod Canal. In 1969, the *Cape Cod Standard-Times* framed the question: "How do you move an old windmill? It is more than 40 feet high and approximately 26 feet in octagonal diameter near the base. Should you cut it in three sections and have them picked up by helicopter? Could it be split down the middle, like a banana, and then transported by flatbed truck? Should it be laid on its side, then hauled to the edge of the canal and lightered across?"

In 1971, contractor John Gallo of Sagamore announced his solution: "We will remove the top, (the sails of the mill already have gone), then a second section just below the top and probably split the lower section vertically in two." [1] It would take three trips to get the windmill from Buttermilk Bay and across the Bourne Bridge to its new Bourne Village location. A new octagonal foundation at the entrance of the Aptucxet Trading Post museum grounds awaited its arrival.

There was just one hitch. The difficulty wasn't in splitting the mill apart, or keeping the pieces from falling off the Bourne Bridge into the canal. The problem was the new foundation: It didn't fit. As the *Sunday Cape Cod Standard-Times* reported, "It was just a pure and simple case of someone having goofed." [2] The already expensive budget of $6,000 needed to be increased by nearly $3,000 to rebuild the foundation. But it was worth it.

Tom Gerhardt showed me through the windmill, which now houses the gift shop of the Historical Society. This unique windmill—well-worth a visit—is open from May through mid-October. Tom is the caretaker of the Aptucxet Trading Post and the Jefferson Windmill. His daughter, Natalina, delightedly searched through the gift shop's toys and souvenirs, as Tom took me to the mill's upper level where Jefferson used to paint. Gerhardt, appropriately, spends much of his time working at the Plimoth Plantation museum and, even in modern clothing, looks like a man of the 17th century.

Landmark House Windmill

10 Hyannis Avenue, corner of Washington Ave., Hyannis Port, Cape Cod, Massachusetts

The Joe Jefferson family compound was, as noted, a predecessor of the Kennedy Compound in Hyannis Port. It was to the latter neighborhood that I next went in search of another eccentric windmill. I had heard of a "tower" on the beach there, but no one I spoke to at Town Hall or at the town's archives knew much about it. Whether it was a water tower or the remains of a windmill, no one seemed to know.

I drove down Ocean Avenue to the corner of Hyannis and Washington Avenues,

Landmark House Windmill, from the veranda, Hyannis Port, MA. Photo by the author.

and there before me, at the edge of the water, was what appeared to be a graceful, pale yellow Dutch windmill without arms. Eight-sided, round-topped, with round portal windows under the eaves, and a gallery, or "stage," around the bottom, it wouldn't have seemed out of place in Kent, England, or Zaandijk, Holland. The stucco exterior matched that of the mansion next to it, a sprawling, European affair with Doric columns and a pergola. A wall surrounded both, and nobody seemed to be home.

I walked onto the beach for a seaside view of the place, then back up to the tower, where I noticed a small sign in one of its lower windows that said "Barber Studio." I was just about to leave, thinking the place had been closed up since winter, when a head popped up above the wall. A kind but cautious voice said, "Can I help you?" This turned out to be the congenial artist Samuel Barber. Sam welcomed me wearing a blue and yellow striped pullover, yellow suspenders, and a bright green bandana tied at his neck.

We entered the mansion, arrayed with Sam's dazzling impressionist paintings. Some were gold-framed, others were unframed and leaning against walls. Sam's paintings are in more than fifty museums and corporate collections, and he's represented by the Walter Findlay Gallery in New York.

I asked Barber how he came to own the Landmark House and mill tower. He and his wife, Janey, saw the mansion in a high wind, just after a terrible winter storm in 1987. It was leaking, it had odd-shaped rooms and that strange windmill tower. But Barber imagined himself painting in a tower by the sea. He immediately wrote a check.

In the late 19th century, George B. Holbrook of Springfield, Massachusetts, built a sprawling, gambrel-roofed house on the site. Holbrook, the treasurer of the

Oil painting of the **Landmark House Windmill** by Samuel Barber. Photo courtesy of Samuel Barber.

American Writing Paper Company in Holyoke, Massachusetts, also built a water tower. This was likely a plain tower with a "pinwheel" windmill to pump water from a well. (Several of these can be seen in early photos of Hyannis Port.) Then, during the Labor Day weekend of 1910, the shingled summer home was completely destroyed by fire.

Early 20th century postcard of Hyannis Port, with windmill pump.

Holbrook was in Western Massachusetts, while his wife and two daughters were still at the summer house. The women had just gone to bed, when the fire erupted from the partitions around the fireplace. Their escape down the main staircase was cut off, so they made their way down a back stairway in their bedclothes. Daughter Emma roused the family chauffeur, who was asleep in his rooms over the garage. A vessel in the harbor saw the fire and sounded its foghorn, which helped awaken the community. Church bells in Hyannis rang to summon the fire department, but the entire house and garage were burned to the ground. The only thing left standing was the water tower.

Barnstable Sheriff Maloney kept guard in the tower to prevent looting until the ashes could be sifted for the family's heirlooms. In the winter of 1911, George Holbook built the large stucco Landmark House now occupying the site, and most likely rebuilt the water tower in the shape of a stucco Dutch windmill.

Barber told me that the place had regularly changed hands every twenty-five years or so since the Holbrooks owned it. The Hunt family of Texas owned the property before selling it to William Gulliver, who in turn sold it to the Horovitz family. Sam bought the place from Betty Horovitz in 1987. The neighbors, including the Kennedy family, took to Sam and his work right away, and his art hangs in many of their homes.

Barber and I exited the house under the pergola and walked the short path to the tower. The first floor was crammed with the paints and canvases of the prolific artist. Sam recalled that when he bought the place, the tower was leaning dangerously and had to be shored up with concrete. It is still a bit cockeyed with a five-inch tilt.

Inspired by the wind and the light, Sam loves painting here, often outdoors on the gallery of the mill tower. As we started to climb the stairway to the second floor, he cautioned me with an anecdote about Ted Kennedy, who knocked his head the first time he went up those stairs.

The most remarkable space in the tower is under its dome. There, Sam has installed a bed suspended by chains so it can sway like a ship captain's bed at sea. Through the round windows of the dome, we could see the Kennedy Compound and Nantucket Sound. Sam mentioned that when John F. Kennedy was a child, he would climb over the sea wall of Landmark House. In later years, Kennedy played football on the beach with Barber's Art Students' League teacher, Bill Draper.

Mostly though, Sam Barber talked about what it was like on summer nights, to sleep in that bed with the stars and water all around...and how those round windows shine out like eyes in a face.

The Wright Windmill

60 Hyannis Avenue, Hyannis Port, Cape Cod, Massachusetts

The search for windmills—so many of them undocumented—involves a good dose of serendipity. I looked everywhere for windmills, and anything vaguely X-shaped (anten-

Elizabeth Roache and daughter, Emma, at the **Wright Windmill**, Hyannis Port, MA. Photo by the author.

nas, tilting telephone poles) would make my head whip around. As I was leaving Sam Barber's place in Hyannis Port, I turned onto Ocean Avenue from Hyannis Avenue. Out of the corner of my eye, I spotted the outline of a round-topped tower peeking up above the surrounding houses. I drove around the block and found my way to 60 Hyannis Avenue, the home of Elizabeth and David Roache. There, I found a graceful, armless, shingled, eight-sided windmill with a white garage door on its first floor.

Liz Roache greeted me at the back door of the large, gabled summer home next to the mill. She and David bought the place in 1995, and Liz says, rather convincingly, that the place was bought for the mill. She and her delightful daughter, Emma, eagerly showed me what Liz calls her "adult tree house."

Emma raced up the stairs of the mill, excitedly pointing out her favorite places. The domed and colorfully painted third floor room was clearly the cherished spot for both Liz and Emma. Liz often uses this room as her studio, where she designs tableware for the American and European markets. Summer dinners often take place in the circular blue room.

Around 1900, the George Wright family of St. Louis built the house and windmill. The windmill, equipped with bathroom and bedroom, housed the Wright's domestic help. Mrs. Wright, in fact, never entered the windmill until the property was sold to the Gardner family. Among the more famous owners of the place was Boston Red Sox star Jim Piersall, whose eight children scrambled through the rambling house and the windmill. Jim and Mary Piersall sold the property to the Paul and Marjorie Yewell family, and the Roaches purchased it from them.

This part of Hyannis Port, previously known as Strawberry Hill, was drawn into the Kennedy circle in the 1950s and early 1960s. John F. Kennedy's sister, Patricia, had married actor Peter Lawford, and the couple rented this house and windmill over a period of three summers during the Camelot years.

As I was leaving, Liz looked at the windmill and confessed the hold it has taken on the Roache family. "When David and I retire," she said, "we'll give the house to Emma, but we'll keep the windmill for ourselves."

Cape Cod Airport Windmill

Race Lane, Marstons Mills, Cape Cod, Massachusetts

The windmill at Cape Cod Airport, Marstons Mills, MA. Photo by the author.

One day, while crossing the Cape through Marstons Mills, I was startled by the sight of a large white windmill across a field of airplanes. I was on the West Barnstable Road (Route 149), when I abruptly turned around to find out why the Cape Cod Airport had a windmill where one might expect to find an air traffic control tower.

I talked to airport manager Dan Lyons, a young man who told me about the Danforths, and the Danforth Family Trust, present owners of the airfield. I had a few questions for the family, but Lyons told me the Danforths were off-Cape and unavailable. As to why an airport should have a windmill, Lyons told me than that Bill Danforth decided to build one for his wife in the 1950s. No other explanation seemed necessary.

I asked Dan what it was like managing a small airport on Cape Cod during the notoriously quiet winters. He said he often played bass guitar in a rock band on weekends—a band sometimes known as "The Pounding Headaches."

I wanted to know, of course, how the windmill fit into all this. It turns out the Cape Cod Airport's resident flight instructor lives in it. Just then, flight instructor John Falvey happened by, so I asked if I might take his photo in front of the mill. He agreed, but said he needed to get something out of the windmill first. A moment later, Falvey came down with a tin of black powder in his left hand. Knowing that black powder is used in historic firearms, I asked if he was a Civil War enthusiast.

Flight Instructor John Falvey and the windmill he lives in, Marstons Mills, MA. Photo by the author.

"No," John laughed. "We're building a potato cannon. We'll drink some beer and fire off some potatoes. It takes a lot of planning. I'm already into this for about $300. The problem, though, is finding potatoes big enough."

I drove away from the airport as the white windmill caught the last rays of light. Pausing to look back, I imagined what the night would look like with planes parked around the windmill and potatoes flying through the air.

Le Petit Moulin

12 Cockachoiset Lane, Osterville, Cape Cod, Massachusetts

Oyster Harbors Windmill

Bridge Street, Osterville, Cape Cod, Massachusetts

Overlooked by most guidebooks, one of the lovelier villages of Barnstable is the out-of-the-way jewel of Osterville, just south of Marstons Mills. Here, I found another little known treasure just by the Osterville Yacht Club and Cosby's Boat Yard, on narrow Cockachoiset Lane. Attached to a lovely shingled house is "Le Petit Moulin," a reproduction windmill built for Boston architect and interior designer Richard Fitzgerald in the 1950s. This is probably one of builder Ivan Kendrick's mills—

Kendrick was the premiere Cape Cod reproduction windmill builder of his time. At the time of my visit, workmen were repainting its interior and making repairs for a new owner. Its previous owner was hooked-rug designer Claire Murray. The new owner is reportedly a millionaire boatbuilder who has competed in the America's Cup race.

In 1925, Forrest W. Norris and a group of investors chose an island in Osterville as one of the most desirable spots on the Cape. On it, they created an exclusive summer colony and named it Oyster Harbors. Nearly two million dollars was put into its facilities—a golf course, tennis courts, stables, roads, water, dredged harbors, and a large clubhouse. Approaching the island on Bridge Street over a causeway, I came to the entrance of Oyster Harbors, looking much as it did in vintage postcards I had collected, with a gatehouse on one side and a reproduction windmill on the other.

Le Petit Moulin, Osterville, MA. Photo by the author.

The island was at one time known as Hannah Screecher's Island, from an old legend that still gets told in town. It's said that Captain Kidd landed on the island seeking a place to bury a chest of gold treasure. A village girl named Hannah, who was hiding in the tangled woods, saw the pirates bury the chest, was taken by surprise and slain. They buried her body in the same pit as the chest. In 1878, Adeline Lovell, a descendant of the first settlers of the island, told the story of Hannah, whose spirit still watches over the treasure:

> *To this day, the faithful Hannah guards it, and moans and shrieks have been heard from that Island which, in time, was called "Hannah Screecher's Island." Men, boys and women have attested to seeing her ghost when they have approached a certain spot in pursuit of berries or wood cutting. Money diggers have left their excavations, which may yet be seen.*[3]

Early 20th century postcard of Oyster Harbors windmill and gatehouse, Osterville, MA.

The name of the island was changed to Oyster Island, then Grand Island, and finally Oyster Harbors, when Forrest W. Norris and other investors bought it in 1925. When the duPont house was built on the island, three skeletons were discovered. One was sent to an institution for study. The other two were stolen from the workmen's camp, leaving no way of knowing whether they were Native Americans or pirates. Or, if one was Hannah, of Screecher's Island.

The Windmill House

Old Mill Point, West Harwich, Cape Cod, Massachusetts

Traveling east along Route 28, I went down Lower County Road in West Harwich. Just where the road crosses the Herring River, there is a splendid view out to Old Mill Point, a scene that looks as much like the broadlands of Norfolk, England, as it does Cape Cod. Since the 1920s, the Windmill House, a steep-roofed, English-style house with a windmill attached, has stood on this point. The mill has a boat-shaped cap, typical of windmills found in Norfolk. The house and mill were built by W.H. Doble, who made his fortune as president of the Quincy Pneumatic Scale Corporation.

Doble bought the land at the mouth of the Herring River, known later as Doble's Point and, still later, as Old Mill Point. Here, he created a private development with a curious collection of architectural styles copied from houses he had seen in Europe—Tudor and Italianate, for example. One house even reproduces a sag he had noticed in the original's slate roof.

Since 1950, the Sugden family has owned the Windmill House. Jane Sugden told me how the place has become a Cape Cod landmark, appearing—without permission—in calendars, postcards, and advertisements. Realty companies have used pho-

Windmill House, West Harwich, MA. Photo by the author.

tos of the windmill to advertise their offices, bringing in a rash of unsolicited offers to buy the property. The windmill, with its three bedrooms and two baths, has been a glorious summer place to raise children. I listened while the Sugden family reminisced about playing in the windmill as children. They would gaze out the small windows, through the windmill's arms, to the surrounding views of the Herring River, the marsh, and Nantucket Sound, or scare the maid by leaping out of a trap door in a closet as she passed from the mill to the dining room.

One of the mill's neighbors recently described what it was like growing up with a windmill in the neighborhood. Writing in the *Cape Cod Voice*, Sarah Korjeff recalled summers at her grandparents' house, when she visited the old mill every day:

> *From a child's perspective, the sloping walls of the windmill were unnatural, but playful. And the few small windows seemed just large enough to reveal someone's face looking out. The idea of having a home with rotating wooden arms, and a roof shaped like a helmet, was quite exciting.*
>
> *There is a house connected to the windmill by a long breezeway. While the house was never the focus of my attention, it too left an impression, with a steeply pitched roof and many angles and additions that contrasted with the rounded shape of the windmill. It all nestled into the bank of the river and was partially hidden from view when you approached from above.*[4]

Early 20th century postcard of **Windmill House**, West Harwich, MA.

When Sarah was a child, her grandparents had shown her a small stuffed donkey that they claimed spent most of his time at the windmill. Korjeff was in her twenties before she realized that animal's name, Donkey Hootie, was a play on words made up by her literary grandmother, a fan of Cervante's famous book.

The Ivan Kendrick and Bill Weintz Reproduction Windmills

Various Locations, Cape Cod, Massachusetts

A Kendrick and Weintz reproduction windmill at the Dolphin Inn of Chatham, MA. Photo by the author.

When I stopped in to see Frank Kennedy at the Dolphin Inn of Chatham, I thought it would be a brief visit. A few words about his small reproduction windmill at the east end of Main Street, and I'd be on my way. Instead, my conversation with Frank set me off in search of other windmills by the Cape's most prolific builders of architectural windmills, Ivan Kendrick and Bill Weintz.

Frank Kennedy's family has owned the Dolphin Inn since 1970. The main inn was built in 1805 by Isaac Hardy. The front half of it was slid over on the ice from Nantucket, and other cottages were added to the property over the years. In the early 1980s, Frank happened to be talking with carpenter Bill Weintz at another Chatham inn. Bill asked Frank if he would like a windmill as a lawn ornament for his inn.

It seems that Bill and his partner Ivan Kendrick had been hired by the town of Brewster to build a windmill as an information booth for the town's Board of Trade. The mill had already been built. Its nine pieces—eight sides and the top—were stacked on a truck awaiting delivery when townspeople realized locating it on the Old Kings Highway violated historic district bylaws. Kendrick and Wientz had a windmill on their hands; thus, it was casually offered to Frank Kennedy for the Dolphin Inn.

Kennedy paid $1,500 for the mill and moved it to Chatham. The interior was originally made of rough-sawn pine held together by boat nails. The blades of the windmill are balanced and turn on an axle from a 1957 Plymouth. The windmill previously held a bedroom, and a garden shed added in 1989 served as a kitchen and bathroom. Today, the windmill is the living room for a larger suite of rooms.

A 16-foot Kendrick windmill, Morris Island, Chatham, MA. Photo by the author.

A 13-foot Kendrick windmill, Morris Island, Chatham, MA.
Photo by the author.

Ivan Kendrick's business was on Crowell Road, where he made sheds for trash cans, as well as axe and shovel handles, until he and Bill Weintz hit upon the idea of building reproduction windmills. They specialized in two sizes: small 13-foot diameter, and 16-foot diameter mills, both with conical caps. These have sprung up all over the mid-Cape. According to son, Ivan, who now owns his father's lumber business, there are at least three mills in Chatham, and others in North Dennis, Cotuit, Osterville, Harwich, and Wellfleet.

I began a tour of these mills by driving across the causeway in Chatham to Morris Island. Here, there are both 16-foot and 13-foot Kendrick windmills on either side of Attucks Trail at the end of the road. After leaving the island, I found other Kendrick windmills in the following days, sometimes entirely by accident.

Sailing into Wellfleet Harbor on a friend's boat one day, I saw a windmill on a bluff to the east of town. I took mental note of it and a few days later bicycled over to have a look. I found it on Sandpiper Hill Road, where I met owners Edith and Richard Gallant. To my surprise, the windmill, added to their house in 1978, was indeed built by Ivan Kendrick.

The Gallant house and its large windmill look down on Wellfleet Harbor and out to Great Island. To the right in the distance are the steeples and narrow streets of Wellfleet. Accepting Edith and Richard's invitation, I found an elegant, pine-paneled dining room on the first floor of their windmill, and a lovely bedroom on its second floor.

The house had begun as a small, rough cottage called "Camp Scott" in 1935. The Gallants purchased it in 1974, and five years later they expanded the house and had Ivan Kendrick build the windmill. After Edith had gone off to run errands, Richard took me through the place and showed me Edith's oil paintings, many of them portraits of the mill. He next took me out to the patio by the windmill where he began to reminisce.

The Gallant house and windmill, Wellfleet, MA. Photo by the author.

Richard had "never heard of the Cape" until he got out of the Navy in 1947. He and a friend drove out for the first time in May of that year from Richard's home town of Acton. Today, Richard is a Cape Codder through and through. He gardens and takes care of the unique problems of owning a windmill—Hurricane Bob, for example, ripped off the mill's blades and tossed them over the roof of the house.

The Gallant family dining room in their windmill, Wellfleet, MA. Photo by the author.

The Robert Horne Windmill

Windmill Lane, Chatham, Cape Cod, Massachusetts

Robert Horne Windmill, Chatham, MA. Photo by the author.

While touring Cape Cod in search of Kendrick and Weintz reproduction windmills, I spied a street sign for Windmill Lane in Chatham. There, I found a small windmill that I at first took to be a 13-foot Kendrick. But rather than a conical cap, it had a more refined, curved boat-shaped one. I contacted the present owners who recalled that the mill had been built by Robert Horne for a development of summer homes in the 1920s. Horne bought the land, laid out plots, and built a windmill which was used as the pump house for the development. The windmill and the house presently connected to it were bought in the 1940s by Robert and Marjenette Hallock, whose son and daughter-in-law, Peter and Peggy Hallock, own it today. Peter remembers sleeping in the mill as a child, listening to the creaking of the wood and the rain pattering on the roof. He now uses the first floor as a workshop, and Peggy paints watercolors in her studio upstairs.

Peggy described summers at the windmill house and the excursions the Hallock family took to their camp on Chatham's outer beach. The entire camp was washed away in the "Perfect Storm" of 1991. Peggy told me of the kerosene lamps and camp beds that washed up on the Chatham shore—debris that the storm also heaped before Barbara Weller's windmill at Sur Mer.

The Elmer Bliss Windmill

Dunroving Ranch, North Road, Chilmark, Martha's Vineyard

The Elmer Bliss Windmill, Martha's Vineyard, MA. Photo courtesy of Julianna Flanders.

Martha's Vineyard has an unusually fine reproduction windmill based on the famous Old Mill that still stands on Nantucket. In 1938, Commodore Elmer J. Bliss had the mill built on his Chilmark property, which he named Dunroving Ranch. It sits on the south side of North Road, near Menemsha.

The Commodore, once known to his friends as the "human amphibian," retired from a life spent half at sea, where he raced schooners in Bermuda, and half on land, where he raised show horses. Commodore Bliss bought hundreds of acres on Martha's Vineyard and began to dream. He had hoped to build a castle-like structure on Prospect Hill to commemorate the Norsemen who he was convinced had visited the island. He planned to call it "Leif Ericson's Lookout."

Other ambitions of the Commodore were easier to realize. He dammed streams, built paddocks and imported ponies from the West. He could be seen roving his ranch, clad in boots, chaps, and a fancy beaded vest. Topped with a ten-gallon sombrero, he must have been a startling sight riding over the Vineyard's rolling hills.

The horizon of Chilmark gained an even more striking sight when Bliss began building his windmill. Not content with a garden-folly mill, the Commodore wanted a fully functioning grain mill—a faithful copy of Nantucket's 18th century Old Mill. Builder and designer Joseph Leonard was hired to draw the plans and build a mill that could grind fine flour. He brought in millstones from Little Compton, Rhode Island, and had them sharpened by a Portuguese mason who worked for him and was familiar with the millstone dressing methods that were still used in windmills in his native Portugal.

On August 16th, 1938, the *Vineyard Gazette* described its construction:

The Elmer Bliss Windmill under construction, Martha's Vineyard, MA. Photo courtesy of Julianna Flanders.

> *The hilltop has been graded, bank-walls have been constructed, and a force of carpenters is at work erecting the windmill, which will stand some forty feet in height. The massive timbers are already prepared, and some have been raised, and let it be mentioned that they are all the old, hand-hewn variety, morticed and tenoned at the ends. The top of the mill will turn, a long timber with a wheel on its end extending from the top to the ground, so that the operator may turn the fan into the wind when it is desired. The fan itself will tower as high again as the mill, and will be equipped with red jute sails, which can be lowered when not in use, and every part will be exact, for the mill will be used when completed.*

The machinery of the mill was made in Leonard's own wood milling shop, but for the exterior he went to the John Duane Wrecking Company in Quincy. According to a letter written by Leonard in 1951, owner Elmer Bliss "was a stickler for the old and the mill just had to look as though it stood there for several hundred years." The long steering boom used to turn the cap of the mill came from the West Coast, but the cartwheel attached to the bottom of the boom was bought from an old local farmer. The windmill's sails were specially made by Gilkey & Durant of New Bedford.

The Tailer Post Mill

Formerly at Windfarm Museum, Martha's Vineyard, presently in Alberta, Canada

Reproducing Nantucket's Old Mill on Martha's Vineyard, as Elmer Bliss did, was a bold idea. What Gerry and Peter Tailer concocted was both visionary and quixotic. Rather than recreate the familiar smock-style mill, the Tailers chose to build the only modern post mill in New England. Traditionally built in England between 1250 and 1840, the unwieldy post mill was squat and balanced on a single, trussed post. The entire mill had to be turned around its post in order to face the wind.

The Tailer Post Mill under construction, Martha's Vineyard, MA. Photo from the collection of William E. Marks.

In the early 1980s, the Tailers decided to build such a mill on their Martha's Vineyard property, and open the Windfarm Museum as a working example of an ancient style of windmill that few living people had ever seen. Peter Tailer wrote of the experience in the Summer 1985 issue of William Marks' *Martha's Vineyard Magazine*:

> *In the spring of 1983 the Tailers completed the two-ton main post by bolting and banding together four 20-foot 15 inch timbers cut and adzed to shape. That summer, while operating the museum, they placed the cross trees on brick piers and then lifted the main post above the cross trees and lowered it into position while placing the supporting quarter bars. It was a very windy day which stressed the rigging to its limits. There was also the threat of heavy rain...Taking turns walking four miles around the capstan, the main post was raised 25 feet in two hours....*
>
> *That fall the Tailers constructed the iron bound crown tree. They gas forged bands and welded up bolts and fittings in their shop.... The crown tree takes the*

> *weight of the entire mill house and allows it to be rotated to face the wind.... While animal fat was a traditionally used windmill grease, the Steamship Authority kindly provided a handful of the extremely heavy lubricant which greases a misaligned ferry into the ramp....*
>
> *In 1984, timber after timber was bolted into place. This past spring, the huge steps, which must be raised to turn the mill, were built and installed. Curved roof rafters were cut out and spiked in place like the inverted ribs of a ship.*

The Tailers had looked forward to installing the windshaft, breakwheel, millstones, and mill arms by 1985. When completed, it would have been one of only four post windmills in the country: two are in Virginia at Williamsburg and nearby Flowerdew Plantation, and a third is in Cambridge, Maryland. The last original American post mill had survived on the Outer Banks of North Carolina but "rotted into oblivion by 1900," according to Peter Tailer.

At Windfarm Museum, the Tailers had hoped to grind varieties of locally grown corn. As Peter Tailer put it, "The great taste advantage of stone ground flour or corn meal is that the stones run cooler than modern steel roller or hammer mills. Thus stone grinding preserves subtle flavors that are destroyed by heat in modern milling processes." Sadly, family illness interrupted their plans. The Tailers had to give up Windfarm Museum before they could complete their windmill.

The unique, eccentric post mill of Windfarm Museum is long gone. I could find it neither on Martha's Vineyard, nor, in fact, anywhere in New England. William Marks offered to track down the missing mill for me, and he recently sent me this message:

"Just received word that the Tailer windmill was shipped to a place in Alberta, Canada, where it is being used to mill grain from farms in the surrounding area." ●

Chapter Six

The Windmill as Whirligig, Lawn Ornament, and Icon of the Miniature Golf Course

"Blake's London in the last two decades of the eighteenth century...
presents us with a woman filling her kettle at the neighborhood pump...
the labourers sitting down with their tankards of porter, the birdcages and pots of flowers
on the windowsills, the shabby man standing on a corner with a sign in his hat saying
'Out of Employ,' a man carrying a plate of pickled cucumber on his head
while another sells toy windmills...."

From Stephen C. Behrendt's review of *Blake: A Biography* by Peter Ackroyd[1]

Windmill at miniature golf course, Sandwich, MA, with lighthouse, whale, and covered bridge in background. Photo by the author.

Travelling through New England looking for windmills, I couldn't help noticing that there are far more imitation windmills on miniature golf courses or lawns than there are real ones. I wondered when we began to be so delighted by windmills that we started reproducing them as garden ornaments and toy whirligigs. We know that Shakespeare was familiar with whirligigs, for he used them as an image in the phrase, "The whirligig of time...." in *Twelfth Night*. And, as noted above, in the 18th century, toy windmills were sold in the streets of William Blake's London.

A painted windmill lawn ornament, Middletown, RI. Photo by the author.

I decided to explore what windmills were out there along the coasts of New England in souvenir and garden shops, and at classic miniature golf courses. In my childhood memory of miniature golf, based largely on numerous games played at the shore in Old Lyme, Connecticut, each course included a covered bridge, a wishing well, a little church, and a windmill. Especially a windmill. Entire afternoons seemed to pass as I attempted to whack my ball between the revolving arms on the windmill hole. This was long before the latest trend in course design... "adventure golf."

I looked in vain for windmills at golf courses like Thunder Falls, Lightning Falls, Cape Escape Mini-Golf (with "Live Koy Fish"), Putters Paradise (with a life-size spouting white whale in a lake), and Pirates Cove ("Visit our haunted caves and dungeons, our brand new 'Pirate Sea Battles' and view our new catapulting 15 foot great white shark!") The unifying theme seemed to be a combination of the Deluge and Disney.

At Thunder Falls Adventure Golf in Hyannis, I met manager Bill White who was busy grooming the course's lawns and trees. When I told Bill I was having trouble finding windmills on mini-golf courses, he looked a little wistful and said, "There are none. The fashion now is water." Bill also cares for the Main Street Mini-Golf in Hyannis, which was built in about 1958. Bill told me they try to preserve the old style of mini-golf at Main Street—with its quaint wooden features like lighthouses and wishing wells—but its windmill kept getting vandalized. Its arms were an easy target, and he finally had to remove the mill. It was replaced by a bear carved out of a tree trunk and a large, indestructible ship's anchor. "That bear," he said, "never gets vandalized. All I need to do is polyurethane it once a year." Bill couldn't let go entirely of the courses' old windmills, though. He keeps two broken down ones in storage.

Contrary to Bill White's impression that mini-golf windmills were extinct, the king of miniature golf windmills came into view at Cape Cod Storyland as I rounded a bend when leaving Hyannis. No ordinary plywood affair, this mill is a detailed scale model. Todd Fedele, whose father owns all three mini-golf sites in Hyannis, took me onto the course. Playing eighteen holes at Cape Cod Storyland takes the golfer through a miniature history of the Cape. The first hole features a reproduction of Plymouth Rock; other holes have historic buildings (the Hyannis West Parish Meeting House and Nantucket's Brant Point Lighthouse), the famous pirate ship *Whydah*, and a Coast Guard Life Saving Station.

Todd Fedele and the Cape Cod Storyland windmill, Hyannis, MA. Photo by the author.

At the sixteenth hole, Todd proudly showed me a reproduction of the historic mill at Eastham. It was built about ten years ago, and since then Todd has reshingled it and kept it in good repair. I asked him if the blades turn. Todd said they are capable of turning but are kept stationary, "Otherwise, they'll clock someone on the head!" Fascinated by the many uses for windmills these days—art studios, gift shops, etc.—I was saddened to hear what Todd finds when he opens up the windmill each spring: empty liquor bottles and cigarette packs, left possibly by teens, and evidence that the homeless may have slept in his windmill on cold nights.

Miniature golf has been a popular American game for over a century. It originated as "Garden Golf," played on real grass with real putters. By the 1920s, bumpers and rails were added and the putting surface was levelled. During the Depression, a particularly wacky form called "rinkie dink golf" was created, which added everything *including* the kitchen sink as obstacles.

The heyday of classic miniature golf lasted from the 1950s through the 1970s. The World Mini Golf Sport Association and its affiliate, the ProMiniGolf Association, lent a certain respectability to the pastime. Windmill holes proved to be the most popular, with their built-in challenge to precisely time putts to get golf balls through the revolving blades. Todd Fedele says the windmill is now an endangered element of the golf course. He blames vandalism and changing taste: the Disney World and "blockbuster action film" effect. Oddly enough, the emphasis on skill has lessened.

Obstacles like windmills are gone; instead, simple putting greens are surrounded by water courses, waterfalls, and elaborate adventure settings. One of the best examples of this trend is the Tree House Adventure Golf course in South Yarmouth. I visited owner Louis Nickinello, who talked to me about the evolution of his family-run business. It began as a classic miniature golf course in 1960 and kept its windmill for thirty-nine years, even after adopting a theme, "The Freedom Trail," in 1984.

In the 1990s, when other courses went for in-your-face adventure themes, Nickinello looked for a new direction for his golf course. His grandchildren came up with the playful idea of the Swiss Family Robinson. The new course—Tree House Adventure Golf—was built in Arkansas. Its original design had a ship's mast as a focal point, but Nickinello wanted something that more clearly spoke Cape Cod—a windmill. A traditional Cape Cod windmill would be out of place, but a water pumping windmill fit perfectly with the Swiss Family Robinson theme. Thus, when it opened in 2001, the new family golf course included, once again, a form of windmill.

In the smaller towns of New England, however, we can still find old miniature golf windmills. I easily found two within a few miles of each other on Cape Cod. The

Tree House Adventure Golf, South Yarmouth, MA. Photo by the author.

The miniature golf course at the Dairy Bar & Grill and Wellfleet Drive-In Theatre, Wellfleet, MA. Photo by the author.

course at Wellfleet's Dairy Bar & Grill on Route 6 has one. It's on the same property as the Wellfleet Drive-In Theatre; built in 1957, it's the only drive-in left on the Cape. A brochure at the Dairy Bar & Grill invites us to "Play a round of retro mini golf on our well-maintained vintage mini golf course. This 18-hole golf course is still the original course built in 1961. Your friends and family will love this bit of ol' Cape Cod." Near its windmill is a wooden 1960s version of a rocket ship.

Down the road at Poit's Family Restaurant and Mini-Golf in North Eastham is a good example of a plywood windmill. Dave Poitras is the third generation owner of the place. His family built the golf course in 1954, and it was Dave's father, Norman Poitras, who constructed the windmill. Set among a wishing well, bridge, and large figures of Porky Pig and Uncle Wiggily, this simple mill has a small motor that moves its arms.

Poit's Mini-Golf, North Eastham, MA. Photo by the author.

Both of the above windmill obstacles are flawed, however, according to serious miniature golf aficionados. Reviews of miniature golf courses can be found on the Internet that measure "Difficulty, Creativity, and Atmosphere" and critique each hole. A review of Connecticut Golf Land Miniature Golf in Talcottville highlights the one thing that really irks mini-golfers about some windmills: "The windmill's blades don't reach ball high, so you can hit it underneath, no matter where the blade is. We found this same problem at other courses we played...."[2] The standards of modern life continue to fall. I went in search of the plethora of windmill souvenirs I used to see in every Cape Cod gift shop when I was a child—whirligigs, lawn ornaments, postcards, soap dishes, ashtrays, placemats, salt and pepper shakers, thimbles, ceramic doorknobs and teapots. Windmill Crafters on Route 28 in South Yarmouth seemed a likely spot: The centerpiece of the building is a small reproduction windmill. Built in the 1950s, and originally located across town, it had once served as a real estate office.

Windmill Crafters, South Yarmouth, MA. Photo by the author.

I met Windmill Crafters' manager, Jill Marjerison, a thirteenth generation Cape Codder descended from Captain Joshua Taylor. Looking around the shop, I couldn't find anything in the shape of a windmill or with a windmill illustration on it. Jill was as disappointed as I was. She told me how hard it was to find such things among American distributors, even though "a lot of people ask us for figurines and magnets of windmills."

On the Lower Cape, at Wellfleet, I stopped at the $2.00 T-Shirt Gift Shop, a sprawling souvenir shop overflowing with seemingly every form of seaside memorabilia. I wandered among the miniature lobster pots, plastic lobsters, lighthouse Christmas tree ornaments, and crab-shaped harmonicas looking for windmills. I found only one: a windmill wind chime. Hung near the ceiling, it had to be fetched for me with a pole.

I asked owner Fred Riley what happened to the windmill souvenirs of my youth. He looked around at dozens of lighthouse items and said, "We're living in the time of the lighthouse. I'll tell you when lighthouses came in—when they started moving them. A few years ago, they moved the one in Truro [back from the eroding cliffs]. They raised a lot of money. Then windmills took a back seat. Now everyone wants a lighthouse souvenir." He seemed to be conjuring a windmill's turning blades when he added, "But these things move in circles—one day windmills are up and lighthouses

Fred Riley holds a windmill wind chime at the $2.00 T-Shirt Gift Shop, Wellfleet, MA. Photo by the author.

are down. The next day lighthouses are up and windmills are down. It's all a cycle. The windmill will have its turn again."

Driving back through Eastham, I looked again at the magnificent windmill on the town green. Just next to it, on Depot Road, I noticed a house called Windmill View. Owned by antiques dealer Dawn Carlson, the property had the best—and most tasteful—collection of windmill lawn ornaments I'd yet seen. Carlson has collected vintage windmills and whirligigs, and added examples of modern stained-glass depictions.

A selection of vintage and contemporary windmill ornaments at Dawn Carlson's home, Windmill View, Eastham, MA. Photos by the author.

Windmill View is an engaging illustration of how the American yard has come to express a family's self-image, dreams, aspirations, and even religious faith. By the mid-20th century, lawns displaying a statue of the Virgin Mary, a lawn jockey, a lighthouse, a windmill, or any combination thereof, were emblematic of the American family in the new suburbia. Built of plastic or wood, the windmill in particular served as a nos-

Pink Flamingos and Plastic Windmills:
A Brief History of Lawn Ornaments

A plastic windmill lawn ornament, Windmill View, Eastham, MA. Photo by the author.

In my travels in New England, I found miniature windmills on lawns from Maine to southern Connecticut, regardless of whether or not an actual windmill had ever existed in the vicinity. I asked myself, "Where did the idea of placing a miniature windmill on a lawn come from?" My research began with an article written by Jennifer Price and published in *The American Scholar* in the spring of 1999. It's an erudite piece titled, "The Plastic Pink Flamingo: A Natural History." In it, Price traces the peculiar history of lawn ornamentation in America.

In the 18th and 19th centuries, American tastes turned against the formality of European gardens. Under the influence of thinkers like Thomas Jefferson and Henry David Thoreau, American landscape architect Lancelot "Capability" Brown preferred natural expanses of meadows, trees, and lakes to the geometric parterres of French gardens. Jefferson's pastoral ideal and Thoreau's faith in the wilderness inclined Americans toward real sheep and native deer over statues of satyrs, Greek gods, and swans.

In the 1840s, nurseryman and designer Andrew Jackson Downing remade country estates into gently sculpted natural landscapes of rolling meadows, streams, and perfectly placed groves. One of his students, Frederick Law Olmsted, became the legendary designer of New York's Central Park. In the 1880s, another student, Frank Jesup Scott, turned to medium-sized lawns and wrote a do-it-yourself manual, *The Art of Beautifying Home Grounds of Small Extent.* He advised homeowners to "avoid spotting your lawn with...plaster or marble images of any kind, or those lilliputian caricatures."

According to Jennifer Price, "The homeowners, however, were getting restless. In the 1920s they bought cast-aluminum animals. In the 1930s do-it-yourselfers made deer, rabbits, and frogs out of cement, the Depression material of choice."

The aftermath of World War II brought the Baby Boom, the suburbs, and plastics. The lawn became a place of leisure and refuge, a place for a barbecue, a swingset, and a variety of fanciful ornaments. The Ben Franklin of lawn ornaments was Don Featherstone. Like Franklin crossing lightning with a kite, Featherstone crossed American lawns and plastic.

In 1956, Featherstone was hired by Union Products in Leominster, Massachusetts, the city that proudly advertizes itself as "America's Plastic City." A year later, Featherstone designed the mother of all lawn ornaments, the plastic pink flamingo. Why the flamingo? New interstate highways in the 1950s made trips to Florida possible for the middle class. A pink flamingo in the yard said that you had been to the Sunshine State, or at least, would like to go there.

Ron Mann sells windmill lawn ornaments at the Mill Store, West Yarmouth, MA. Photo by the author.

talgic symbol of both old Europe and old Cape Cod. And it had the added attraction of being a wind-propelled action toy.

Toy windmills, or at least rudimentary whirligigs, date back into the mists of the history of toys. They were familiar, as we've seen, to both Shakespeare and William Blake. In the 20th century, it was Chatham writer Joseph Lincoln who gave the modern toy windmill industry a kick-start. In his 1918 book, *Shavings: A Novel*, Lincoln featured a toy windmill maker named J. Edgar W. Winslow. Such was Lincoln's influence in the creation of mythical Cape Cod, that what began as literally a cottage craft turned into a craze. Toy windmills soon spun on seemingly every fencepost and porch on the Cape and throughout much of New England.

The Shavings Shop, Chatham, MA, an inspiration for Joseph Lincoln's book, *Shavings*. Photo courtesy of the Chatham Historical Society.

In the first chapter of *Shavings*, Lincoln describes Gabriel Bearse on his way to the post office in fictional Orham—a combination of Orleans and Chatham. Bearse spies a toy windmill, fastened to a white picket fence, "chattering cheerfully as its arms spun in the brisk, pleasant summer breeze." Lincoln continues:

> *The little windmill was one of a dozen, all fastened to the top rail of that fence and all whirling. Behind the fence, on posts, were other and larger windmills; behind these, others larger still. Interspersed among the mills were little wooden sailors swinging paddles; weather vanes in the shapes of wooden whales, swordfish, ducks, crows, seagulls; circles of little wooden profile sailboats, made to chase each other 'round and 'round a central post. All of these were painted in gay colors, or in black and white, and all were in motion. The mills spun, the boats sailed 'round and 'round, the sailors did vigorous Indian club exercises with*

their paddles. The grass in the little yard and the tall hollyhocks in the beds at its sides swayed and bowed and nodded. Beyond, seen over the edge of the bluff and stretching to the horizon, the blue and white waves leaped and danced and sparkled. As a picture of movement and color and joyful bustle the scene was inspiring; children, viewing it for the first time, almost invariably danced and waved their arms in sympathy. Summer visitors, loitering idly by, suddenly became fired with the desire to set about doing something, something energetic....

At the rear of the little yard, and situated perhaps fifty feet from the edge of the high sand bluff leading down precipitously to the beach, was a shingled building, white-washed, and with a door, painted green, and four windows on the side toward the road. A clamshell walk led from the gate to the doors. Over the door was a sign, very neatly lettered as follows: "J. EDGAR W. WINSLOW. MILLS FOR SALE."

The original role models for Lincoln's fictional toy windmill maker can be found among the early postcards and photographs of the Cape. There were Baker's Windmill Shop in West Dennis, and Atherton Crowell's Toyland in Dennis Port. Edward Chase, a Navy veteran, chauffeur, and carpenter, recalls that as a boy he made $1.00 a day—for a 14-hour day—carving miniature windmills on a bandsaw at Crowell's Toyland.

Baker's Windmill Shop, West Dennis, MA.

Joseph Lincoln was well aware that the toy windmill was hardly a Cape Cod invention. In his 1935 book, *Cape Cod Yesterdays*, he refers to their ancient origins and recounts the accepted version of the beginnings of Cape Cod's unique variation:

> *Some twenty or more years ago* {before 1915}, *so they tell us, a surfman attached to one of the Cape Cod lifesaving stations, having spare time on his hands and being "handy with tools," decided to make some toys for his own children. Windmills attached to the end of a stick had been made for centuries, but this surfman's ambition was for something more pretentious. So his little mills were miniature replicas, so far as their exterior appearance was concerned, of the real mills which he could see upon the tops of the hills over on the mainland.*
>
> *His children played with them and mounted them on posts in the front yard. The neighbors saw them and mounted others for their children. Summer visitors saw them and begged the opportunity to buy or to order. And, before long, that lifesaver retired from the service and took up toy windmill making as a business. His success bred imitators and now each town from, and including, Falmouth to Provincetown, makes and displays yards full of wooden toys....*
>
> *They are made and shipped all over the country. We have seen "Genuine Cape Cod Windmills" displayed and offered for sale in Florida and in California.*

Atherton Crowell's Toyland, Dennis Port, MA.

Seabury Surf Shop, with modern whirligigs, Wellfleet, MA. Photo by the author.

As physical objects, windmills are irresistible. We love to see the graceful lines of authentic old mills, but we also love the windmill's comic side. In addition to producing flour and energy, the windmill has served as a jungle gym for brave children, a plastic whirligig, painted birdhouse, refrigerator magnet, and mini-golf obstacle. Useful, graceful, playful, and sometime cheesy, the windmill has proved to be the most versatile product of industrial evolution. The windmill, in its many forms, can stand proudly next to a nuclear reactor...or a pink plastic flamingo. ●

Ceramic windmill teapot at a Cape Cod antique shop. Photo by the author.

Chapter Seven

The Windmill in Our Imagination: *Literature, Music, Poetry, and Film*

"But look, your Grace, those are not giants but windmills,
and what appear to be arms are their wings which, when whirled
in the breeze, cause the millstone to go."

Miguel de Cervantes, *Don Quixote*

Illustration from *Mr. Wind and Madam Rain*, a book of Breton folk stories by Paul de Musset, Harpers & Brothers, n.d.

The windmill has been broadly and deeply embedded in our creative lives since it first turned a pair of millstones. We see the windmill as an eternal connection to nature, an elemental symbol of wind and weather. Because they furnished the necessities to sustain our ancestors, windmills remain emblems of the bread and salt of life, symbols of fertility and the harvest. In windmills, poets saw more than the blowing of wind and turning of wheels. In them they saw the spinning movements—sometimes stately, sometimes out of control—of the stars and planets.

Miguel de Cervantes created the most enduring windmill image in 1605. He forever tied windmills to his character Don Quixote, the visionary madman with whom we somehow identify. Quixote fantasized giants where there were windmills and jousted with them. He imagined armies where there were sheep, and a fair lady where there was none. Quixote freed criminals, envisioning he was helping the oppressed. He thought of himself as a knight, but the people around him teased and tormented the poor old man.

Cervantes' most famous chapter is titled, "Of the good fortune which the valorous Don Quixote had in the terrifying and never-before-imagined adventure of the windmills, along with other events that deserve to be suitably recorded." Quixote has just convinced his neighbor, Sancho Panza, a poor farmer "with few wits in his head," to become his squire when they confront their first ordeal:

> *At this point they caught sight of thirty or forty windmills which were standing on the plain there, and no sooner had Don Quixote laid eyes upon them than he turned to his squire and said, "Fortune is guiding our affairs better than we could have wished; for you see there before you, friend Sancho Panza, some thirty or more lawless giants with whom I mean to do battle. I shall deprive them of their lives, and with the spoils from this encounter we shall begin to enrich ourselves; for this is righteous warfare, and it is a great service to God to remove so accursed a breed from the face of the earth."*
>
> *"What giants?" said Sancho Panza.*
>
> *"Those that you see there," replied his master, "those with the long arms some of which are as much as two leagues in length."*
>
> *"But look, your Grace, those are not giants but windmills, and what appear to be arms are their wings which, when whirled in the breeze, cause the millstone to go."*
>
> *"It is plain to be seen," said Don Quixote, "that you have had little experience in this matter of adventures. If you are afraid, go off to one side and say your prayers while I am engaging them in fierce, unequal combat."*
>
> *Saying this, he gave spurs to his steed Rocinante, without paying any heed to Sancho's warning that these were truly windmills and not giants that he was riding forth to attack. Nor even when he was close upon them did he perceive what they really were, but shouted at the top of his lungs, "Do not seek to flee, cowards and vile creatures that you are, for it is but a single knight with whom you have to deal!"*
>
> *At that moment a little wind came up and the big wings began turning.*
>
> *"Though you flourish as many arms as did the giant Briareus," said Don Quixote when he perceived this, "you still shall have to answer to me."*
>
> *He thereupon commended himself with all his heart to his lady Dulcinea,*

beseeching her to succor him in this peril; and, being well covered with his shield and with his lance at rest, he bore down upon them at a full gallop and fell upon the first mill that stood in his way, giving thrust at the wing, which was whirling at such a speed that his lance was broken into bits and both horse and horseman went rolling over the plain, very much battered indeed. Sancho upon his donkey came hurrying to his master's assistance as fast as he could, but when he reached the spot, the knight was unable to move, so great was the shock with which he and Rocinante had hit the ground.

"God help us!" exclaimed Sancho, "did I not tell your Grace to look well, that those were nothing but windmills, a fact which no one could fail to see unless he had other mills of the same sort in his head?"

"Be quiet, friend Sancho," said Don Quixote. "Such are the fortunes of war, which more than any other are subject to constant change. What is more, when I come to think of it, I am sure that this must be the work of that magician Freston, the one who robbed me of my study and my books, and who has thus changed the giants into windmills in order to deprive me of the glory of overcoming them, so great is the enmity that he bears me; but in the end his evil arts shall not prevail against this trusty sword of mine."[1]

The millstone itself has been an abiding image and symbol. Ironically, the stones which nourish life have more often been used as images of burden and sin than of bounty. The metaphor of having a millstone around one's neck dates back to the Bible: "Things that cause people to sin are bound to come, but woe to that person through whom they come. It would be better for him to be thrown into the sea with a millstone tied around his neck than for him to cause one of these little ones to sin." (Luke 17:1-3, *The Holy Bible: New International Version*)

In the Bible, the millstone served as both a weapon, as in the death of Abimelech (Judg. 9:53), and a comfort. The sound of grinding millstones was familiar throughout the cities and towns of Palestine. Small village populations used hand mills, while larger cities had community mills turned by slaves or animals. The sound of grinding was an assurance that life continued. When it was no longer heard, it signified complete desolation (Jer. 25:10; Rev. 18:22).

As ancient cultures developed wind power, they invariably sought among their gods for the origins and meaning of such power. In *The Dictionary of Symbolism*, Hans Bjerdermann, explains:

For the symbologist, winds are not merely currents of air but also supernatural manifestations of divine intentions. Two characteristics are of primary importance: the unpredictability of the wind, and its ability to produce dramatic effects despite its own invisibility. Where winds come from a characteristic direction (like the sirocco or the bora), they are easily personified, as in Greek antiq-

> *uity. The bitter north wind Boreas abducts the Athenian princess Oreithyia and carries her off to his home in Thrace; Zephyrus, the mild west wind, brings the young Psyche to Eros, the god of love....*
>
> *In ancient China the wind (feng) was originally revered as a bird god... Feng-shui is the science of "wind and water," the geomantic choice of locations for buildings on the basis of natural features of the landscape. Feng also has extended meanings: "caressing" and "odor." A fortune-teller is called a "mirror of the winds."*
>
> *In ancient Iran, as in Islam, the wind was thought of as a great organizing principle for the cosmos. In ancient Egypt the cooling north wind came from the throat of the god Amon, and the name of the Sumerian god Enlil literally means "puff of wind"....*
>
> *In ancient Mexico the wind (ehecatl) was associated with the god Quetzalcoatl, who in this context wears a beak-like mask over his face.*[2]

Images of the wind abound in Biblical Scripture. In medieval society, where the extent of literacy was largely confined to the Bible, these images were naturally associated in the popular mind with windmills. The wind was a token of God's unseen power and judgment (Hos. 13:15). The wind was the Lord's instrument in overcoming chaos (Gen. 1:2, 8:1), and the means by which the prophets were transported (I Kings 18:12; II Kings 2:16). When the Lord was revealed to the prophet Elijah on Mount Horeb, it was partly through the sign of the wind: "And, behold, the Lord passed by, and a great and strong wind rent the mountains, and brake in pieces the rocks before the Lord; but the Lord was not in the wind...." (I Kings 19:11).

Sometimes the wind was simply inscrutable: "The wind blows where it wills, and you can hear the sound of it, but you do not know whence it comes or whither it goes...." (John, 3:8; Eph. 4:14). From the Renaissance through the 19th century, this sense of the mystery of the wind endured. In Shakespeare's *The Comedy of Errors*, we read, "There's something in the wind" (Act III, Scene 2). And in *The Infanta's Dream*, Victor Hugo wrote, "Madame, bear in mind that princes govern all things—save the wind."

In the 19th century, Henry David Thoreau's eager, open mind surveyed the span of Eastern and Western religion for meaning. He often wondered what significance there might be in the wind. "The wind roars amid the pines like the surf. You can hardly hear the crickets for the din.... Such a blowing, stirring, bustling day,—what does it mean?.... The whole country is a seashore, and the wind is the surf that breaks on it." [3]

Thoreau distilled his thoughts on the wind into one of his best poems:

> *Men say they know many things;*
> *But lo! They have taken wings,—*

The arts and sciences,
And the thousand appliances;
The wind that blows
Is all that any body knows.[4]

The 20th century brought two World Wars, existentialism, the presumed death of God, and the "Age of Anxiety." Thoreau's line, "The wind that blows/ Is all that any body knows," anticipated the new century. W.H. Auden, one of its central poets, used the winds as a trope for his age in "If I Could Tell You":

The winds must come from somewhere when they blow,
There must be reasons why the leaves decay;
Time will say nothing but I told you so....[5]

In a sense, Auden was followed in spirit by Bob Dylan. Just as Auden reflected the anxieties of his era, Dylan wrote in the 1960s of the anxieties of his generation. Searching for the meaning behind racism, alienation, and the nuclear arms race, Dylan's refrain was, "The answer is blowin' in the wind."

Windmills receive the benefit of the mysterious wind and are physical proof of its movement. Thus, they have assumed many of the wind's mythical traits. Because the wind has traditionally been associated with the power of gods, fate, chaos, destruction, and madness, the windmill has come to share in this symbolism, as well.[6] Because windmills also sustain life—through production of bread and salt—the pumping of water, and the modern production of energy, they are also emblems of goodness and bounty.

Creative uses of windmill imagery appear in the works of the world's greatest dramatist. Shakespeare used one as a symbol of humble laboring life in *Henry IV, Part I*: "I [Hotspur] had rather live/ With cheese and garlic in a windmill, far,/ Than feed on Cakes and have him [Owen Glendower] talk to me/ In any summer house in Christendom" (Act III, Scene 1, 158). Shakespeare managed to use the windmill, in the form of a whirligig, in a wild speech given by the Clown in *Twelfth Night*. Feigning madness, the Clown comments on the follies of the other characters. He ends his rant with the profoundly fatalistic line, "And thus the whirligig of time brings in his revenges" (Act V, Scene 1, 360-366).

In the 17th century, French fairy tale writer Charles Perrault used a windmill as the background for his story of *Puss in Boots*. It begins, "A certain miller had three sons, and when he died the sole worldly goods which he bequeathed to them were his mill, his ass, and his cat.... The eldest son took the mill, and the second son took the ass. Consequently all that remained for the youngest son was the cat, and he was not a little disappointed at receiving such a miserable portion." [7] The windmill embodied security and survival, and the ass was a means by which the second son might earn a living; but the cat appeared capable only of catching mice while hiding amongst the meal. However, by the end of the story this cat cleverly acquires riches for the miller's third son and arranges a marriage between him and the king's daughter.

The 19th century was the last to use wooden windmills before the Industrial Revolution rendered them obsolete. The windmill continued to appear in many forms of literary allusion until the latter years of the 1800s, when it began its career as a nostalgic emblem of a lost agrarian world.

Interestingly, Leo Tolstoy used a toy windmill in *Anna Karenina*. Here, the windmill is an object of imagination and distraction for the young Seryozha. The boy loses himself, and forgets his mother for a while, by imagining himself whirling with the arms of the mill:

> *Vasily Lukich was in good humor, and showed him* [Seryozha] *how to make windmills. The whole evening passed in this work and in dreaming of how to make a windmill on which he could turn himself—clutching at the sails or tying himself on and whirling around. Seryozha did not think of his mother the whole evening, but when he had gone to bed, he suddenly remembered her, and prayed in his own words that she would stop hiding herself and come to him tomorrow for his birthday.*[8]

In *Great Expectations* (1861), Charles Dickens employs the dark, forlorn aspects of windmills in a passage describing the loneliness and despair of young Pip:

> *It was wretched weather; stormy and wet, stormy and wet; and mud, mud, mud, deep in all the streets. Day after day, a vast heavy veil had been driving over London from the East, and it drove still, as if in the East there were an Eternity of cloud and wind. So furious had been the gusts, that high buildings in town had had the lead stripped off their roofs; and in the country, trees had been torn up, and sails of windmills carried away; and gloomy accounts had come in from the coast, of shipwreck and death.*[9]

Edgar Allan Poe's *The Narrative of Arthur Gordon Pym of Nantucket* (1838) contains a scene of men lashed to fragments of a windlass on a ship battered in a severe gale. Pym falls into "a state of partial insensibility," and his mind becomes filled with wild images:

> *I now remember that, in all which passed before my mind's eye,*
> *motion was a predominant idea. Thus, I never fancied any stationary object,*
> *such as a house, a mountain, or anything of that kind; but windmills, ships,*
> *large birds, balloons, people on horseback, carriages driving furiously, and*
> *similar moving objects, presented themselves in endless succession.*[10]

In Herman Melville's exotic novel, *Mardi and a Voyage Thither* (1849), the image of a windmill is used to illustrate the movement of the madman, Alanno, in the central temple of Vivenza: "One hand smiting his hip, and the other his head, the lunatic thus proceeded; roaring like a wild beast, and beating the air like a windmill...." [11]

As the 19th century wore on and the Industrial Revolution flourished, windmills became much more romantic figures. In R.D. Blackmore's *Lorna Doone* (1869), windmills were quaint objects in the landscape. In a passage about his boyhood, the narrator comments, "Perhaps for a boy there is nothing better than a good windmill to shoot at, as I have seen them in flat countries...." [12]

Robert Louis Stevenson romanticized windmills in his poem, "To Nelly Sanchez." In reference to the fantasy and imagination found in books, he writes,

...There, with these,
You shall give ear to breaking seas
And windmills turning in the breeze,
A distant undermined din
Without; and you shall hear within
The blazing and the bickering logs,
The crowing child, the yawning dogs....[13]

John Greenleaf Whittier's poem, "The King's Missive" (1894), contains this reference to Boston's first windmill:

...The autumn haze lay soft and still
On wood and meadow and upland farms;
On the brow of Snow Hill the great windmill
Slowly and lazily swung its arms;
Broad in the sunshine stretched away,
With its capes and islands, the turquoise bay;
And over water and dusk of pines
Blue hills lifted their faint outlines....[14]

Henry Wadsworth Longfellow created the first speaking windmill in literature in "The Windmill." His poem was part of the new romanticization of windmills, while at the same time it referred back to the scene in Cervantes where Don Quixote mistook windmills for giants:

Behold a giant am I
Aloft here in my tower
With my granite jaws I devour
The maize and the wheat and the rye
And grind them into flour.

I look down over the farms
In the fields of grain I see

The harvest that is to be
and I fling to the air my arms
For I know it is all for me.[15]

In the 20th century, windmills began to rapidly disappear. By the 1930s, when Cape Cod's oldest windmill was moved to a museum in Michigan, the romance of the windmill turned to full-blown nostalgia. In 1934, Isaac Small published a typical lament, called "The Old Mill." He addressed it to the Truro windmill on Cape Cod, which had been torn down in 1872:

A hundred years and more have passed,
Since workmen reared thy form;
And bade thy sturdy frame of oak,
Defy the wind and storm.

How many hours thy swinging arms
Have faced the northern wind,
And moved thy creaking wheels along,
The shining grain to grind.

The child with wondering gaze hath stood
By thy great oaken door,
And watched the miller tend the grist,
Or scan the clothes he wore.

With what delight the child hath leaped
Upon the bench within
And caught the warm grain in his hand
That filled the wood bin.

The whirling wheels with many cogs;
The great hewn beams o'erhead,
The grinding stones, the hopper box,
Through which the soft grain sped.

The cobwebs from the rafters hung;
The floor was deep and strong,
The miller in his whitened coat
Lightened his work with song.

But rain and snow and many gales,
And want of tender care;
The burning sun, and father time
Has stripped thy timbers bare.

No more the youth shall play beside,
Nor maiden tell her story,
The Grand old Mill no more will turn,
Gone are its days of glory.

No more its mighty arms shall swing,
Nor winding stair the miller climb,
Its wheels are stopped, its race is run,
Cut down by pruning hook of time.

Your stout frame stood the shock of storms;
How sad to see you fall,
But fate like yours is only what
Must some day come to all.

Thy crumbling ruins lie
About the grassy hill,
With saddened hearts we now must say
A last good bye OLD MILL.[16]

Other 20th century writers and musicians went on to make bold, creative, sometimes humorous use of the windmill. In Eudora Welty's short story, "The Worn Path," Phoenix Jackson witnesses the transformation of Southern blacks from slave to citizen after the Civil War. Phoenix dreams of the day her grandson can go to college and buys him "a little windmill they sells, made out of paper. He going to find it hard to believe there is such a thing in the world."

Kingsley Amis took a droll look at the wind and windmills in his poem, "Ode to the East-North-East-by-East Wind":

We know, of course, you blow the windmills round,
And that's a splendid thing to do;
Sometimes you pump up water from the ground;
Why, darling, that's just fine of you![17]

Poet Cate Marvin also finds lightness in the theme. In "A Windmill Makes a Statement," she describes a love scene between the wind and a stainless steel windmill lawn ornament:

> *...On lawns, on lawns we stand,*
> *we windmills make a statement. We turn air,*
> *churn air, turning always on waiting for your*
> *season. There is no lover more lover than the air....*[18]

In a more brooding vein, the great and tragic poet Sylvia Plath wrote in her journals of "being crushed in a huge dark machine, sucked dry by the grinding indifferent millstones of circumstance." [19] In common with earlier writers, Plath employs the imagery of a "dark machine" and "millstones" to tie the mill to fate and mortality.

In the 20th century, windmills began appearing in decidedly non-literary, non-traditional, and non-agrarian settings. In 1900, a windmill became the symbol of the bawdy side of Parisian nightlife when the city's famous Moulin Rouge was at the height of its popularity. And the windmill later became the emblem of the largest house of prostitution in the world: In the 1930s, the "Molino Rojo" stood above the Tijuana racetrack in Mexico. This shag-carpeted emporium covered half a city block. The building was topped with an enormous spinning windmill, trimmed in flashing red lights. This beckoning windmill could be seen from across the border in California. It was, perhaps, the Moulin Rouge and the Molino Rojo that led to the use of windmills as settings for the lusty, dark, and mysterious sides of life.

Modern best-selling novels and popular songs have used, and misused, the windmill in a variety of ways. Sidney Sheldon's glitzy *The Windmills of the Gods* (made into a TV film starring Jaclyn Smith in 1988) got the working principle of the windmill reversed. Sheldon used a quote from H.L. Dietrich as an epigraph to this bestseller of international intrigue: "We are all victims, Anselmo. Our destinies are decided by a cosmic roll of the dice, the winds of the stars, the vagrant breezes of fortune that blow from the windmills of the gods." Wind, of course, blows *at* windmills, not *from* them.

Windmills were again associated with the wildly turning lives of fictional characters in Vicky Martin's *The Windmill Years* (1978). Like Sidney Sheldon, Martin's characters move in glamorous settings, in this case (as the dust jacket entices), "the glittering, high-powered world of art collecting—a world of fantastic riches, power, and greed."

In the 1960s, lyricist Alan Bergman and composer Michel Legrand wrote "The Windmills of Your Mind." In a rush of words, the popular song likens the "windmills of your mind" to circles, spirals, wheels within wheels, carousels spinning around the moon, and snowballs careening down a mountain. The song captured some of the drug-inspired imagery of the period. Recorded many times, notably as a hit by Dusty Springfield, it spoke of loss, fleeing time, and the edges of desperation.

In the mellower 1970s, Marty Balin of the Jefferson Starship wrote a hit called "Miracles." Unlike the desperate frenzy of the Bergman/Legrand ballad, Balin's love song invoked the simple revery of windmills and rainbows.

In 1992, a quirky new group came on the scene called They Might Be Giants. The group has been described in the press as "Buddy Holly crossed with the Talking Heads, crossed with 'Sesame Street,' crossed with poet Sylvia Plath, crossed with the physics-warping art of M.C. Escher." [20] The group's name derives indirectly from Don Quixote's reaction upon seeing a field of windmills. Group founders came on the phrase, not in Cervantes, but as the title of the 1971 George C. Scott film, *They Might Be Giants*. In the film, Scott plays a retired judge who believes he is really Sherlock Holmes. The film uses Scott's character to make a muddled reference to the madness of Don Quixote.

The darkest use of a windmill in popular music is in the bitter rap song, "Is This the End? (Part Two)," by Puff Daddy. It contains an unfortunate line describing the victim of a murder spinning like a windmill from the force of gunshots. Conversely, the lightest use of windmills in pop music comes from the Hong Kong group called Windmills. The band describes itself on its web site as follows:

> *Our songs are innovative and fresh. The name Windmills evokes images of fresh air, rolling fields, wild flowers and sunny days. It also brings to mind the imagery of Don Quixote tilting at windmills. To fight with things we can see is not hard—we believe that only the things we cannot see are the real monsters. Windmills believes every one should keep a windmill in their heart to keep the mind clear.*[21]

Screenwriters and directors have used the windmill to evoke suspense, horror, serenity, sensuality, and occasionally humor. Windmills appeared, of course, in *Man of La Mancha*, the 1972 film based on the musical version of *Don Quixote*. But perhaps the most dramatic windmill scene in film history is the climax of the classic horror film, *Frankenstein*. In the 1931 film, Boris Karloff appeared as "The Monster" and Colin Clive as Dr. Frankenstein. As the angry townspeople search the countryside, Frankenstein's monster finds shelter in a crumbling windmill. He hurls Dr. Frankenstein into its turning blades, from which the doctor is thrown to the ground. The mob sets its torches to the windmill, which bursts into flames against the dark sky, and the Monster dies.

The dusty gears and creaking arms of a crumbling windmill were used to good effect by Alfred Hitchcock in his 1940 film, *Foreign Correspondent*. Joel McCrea, as the title character, goes to Europe on the eve of World War II. While covering a peace conference in Amsterdam, he witnesses what appears to be the assassination of a Dutch diplomat. The diplomat is, instead, spirited away by the Germans to a windmill. McCrea discovers the kidnapped diplomat in a scene suffused by images of the shadowy gears and stairways of the windmill's interior. While trying to avoid discovery by the Germans, McCrea catches his coat in the mill's brakewheel, then climbs out a window and clings to the cap of the mill. Later, the diplomat's captors use the arms of the windmill to signal their planes, just as actual windmills in Holland signalled Allied planes during World War II.

Filled with both Hitchcock's trademark intensity and humor, the film finds the correspondent the target of a wry accusation: "All you care about," complains his leading lady, "is having fun with windmills and hotel bathrooms."

Director Sam Raimi used an abandoned windmill for a comic scene in his 1992 film, *Army of Darkness.* Bruce Campbell portrays Ash, a man sucked back in time to the year 1300 A.D. Caught between warring armies on horseback, Ash is equipped with a chainsaw, shotgun, and battered automobile—all of which time-travelled with him. While on a quest for a Sumerian book of the dead, Ash spends a night in an old mill which inspires wild hallucinations, during which he splits in two and does battle with himself.

Two films versions of *Moulin Rouge* have been filmed, nearly fifty years apart. John Huston directed the first in 1952. In it, artist Toulouse-Lautrec finds brief solace in the seedy, cognac-drenched dissipation of the Moulin Rouge. Crippled in body and soul, he tries to gas himself in his apartment, then changes his mind and throws open the windows. Lautrec looks out over the rooftops of Paris and sees two windmills in the sun.

The Moulin Rouge, Paris. Photo by Wendy Moss Thomas.

Director Baz Luhrmann's *Moulin Rouge* of 2001 bears little resemblance to the original. Lautrec is replaced by a young, handsome writer, Christian (Ewen

McGregor), who falls in love with the courtesan Satine (Nicole Kidman). Only the Moulin Rouge remains the same, described in the film as "a nightclub, a dance hall and a bordello; a kingdom of night-time pleasures where the rich and powerful come to play with the wretched creatures of the underworld."

The nightclub's windmill is a far more pervasive image in this film, however, than in John Huston's version. Most of the young lovers' scenes of hope and despair are punctuated by the turning of the windmill's light-covered arms. In a dazzling climax, the fantasy world of the Moulin Rouge clashes violently with the real lives of the lovers. Satine's rich sponsor wants her lover dead and sends an armed killer on stage to do the job. Fortunately, Christian survives, but only to find that Satine is dying of tuberculosis.

It appears the grinding windmill of fate will destroy the lovers' dreams. Yet, love is, in the end, triumphant, but at an unbearable cost. Death may separate Christian and Satine, but their love lives beyond death. In one of the final scenes of *Moulin Rouge*, the windmill still turns in a snow-covered Paris. ●

Chapter Eight

Try and Catch the Wind: *The Struggle for Clean Energy*

"Occasionally a new invention will appear which will greatly affect a whole range of allied inventions and industries in such a way as to entirely change time-honored customs, inaugurate new practices and establish new arts."

Scientific American, December 20, 1890

Searsburg Wind Power Facility, Searsburg, VT, 2003. Photo by the author.

Radically new technologies are always met with varying portions of excitement and fear. When Romans began using waterwheels, they looked over their shoulders, certain that the goddess Ceres would punish them for capturing the power she had given to the rivers. Similar anxiety came with the harnessing of wind power in Persia in the 7th century, A.D. Throughout the history of windmills there have been occasional controversies—the Cape Cod windmill that disemboweled a horse in 1829, for example—but the benefits of the technology outweighed the risks of using it. The first windmill generated electricity more than 120 years ago. Why, then, are there more windmill lawn ornaments today than there are wind turbines producing energy?

In the winter of 1887-1888, Charles Brush was busy building the world's largest windmill. Brush had already developed the first commercial arc light and a very efficient DC dynamo for the public electric grid. His company, Brush Electric, of Cleveland, Ohio, had even come up with an efficient way of making lead-acid batteries. Now, in the backyard of his mansion on Euclid Avenue, the inventor was connecting all these elements to create the first automatically operated, power-producing windmill.

Charles Brush's electric power windmill, 1887-1888. Photo courtesy of the Danish Wind Industry Association.

Brush's mill was massive, with 144 rotor blades made of cedar spreading 50 feet in diameter. For twenty years, the windmill charged batteries and powered lights for his magnificent mansion.

Brush's backyard windmill didn't catch on, however, because it was far too inefficient. In an article published on December 20, 1890, *Scientific American* magazine noted of his mill: "Despite the size of the turbine, the generator was only a 12 [kilowatt] model. This is due to the fact that slowly rotating wind turbines of the American wind rose type [shaped like a pinwheel] do not have a particularly high average efficiency." The journal added this caveat: "The reader must not suppose that electric lighting by means of power supplied in this way is cheap because the wind costs nothing. On the contrary, the cost of the plant is so great as to more than offset the cheapness of the motive power. However, there is a great satisfaction in making use of one of nature's most unruly motive agents." [1]

Brush, considered one of the founders of the American electricity industry, sold Brush Electric in 1889. The company merged with Edison General Electric Company in 1892, under the name General Electric. But it was up to the Danes, and particularly an inventor named Poul la Cour, to take wind energy to the next level.

Denmark had been developing wind-generated electricity around the same time as Charles Brush, and the first Danish prototypes began appearing by 1890. Pioneer Poul la Cour approached the problem from the discipline of aerodynamics. Using his own wind tunnel, he discovered that fast rotating wind turbines with few rotor blades were more efficient for electricity production than slow moving wind turbines.

La Cour's experiments using both wind power and hydrogen had literally explosive results. He used wind to produce electricity, then used the electricity to produce hydrogen to fuel gas lights. La Cour tried this out at the Askov Folk High School, where he taught budding young wind electricians, but he abandoned this particular approach after hydrogen explosions blew out the windows in several of the school's buildings—several times.

In 1904, la Cour founded the Society of Wind Electricians which grew to 356 members in its first year. His *Journal of Wind Electricity* was the world's first publication of this type. Because of la Cour, Denmark presently leads the world in the field. By 1918, Denmark had 120 wind turbines powering local utilities. This accounted for 3 percent of Danish electricity at that time.

Wind electricians graduating from la Cour's class at Askov Folk High School, Askov, Denmark. Photo courtesy of the Danish Wind Industry Association.

The use of windmills—for all purposes—declined steadily in the industrial age. In Europe, the steam engine replaced water-pumping windmills, and in the U.S., the Rural Electrification Administration brought inexpensive electric power to rural areas in the 1930s. Cheap coal and oil prices halted the development of wind power until World War II, when a supply crisis occurred. Only then were large windmills again seen as an alternative energy source.

A few years prior to the war, in 1934, Palmer Cosslett Putnam built a home on Cape Cod. Finding both the wind and his electric rates high, the young engineer, aviator, and yachtsman, came up with an idea that's widely endorsed today. He wanted to build a windmill that would power his home and feed power back to the electric company when the wind was excessive. Realizing the technology for his bold plan didn't exist, Putnam dreamed even larger. He enlisted Thomas S. Knight, fellow yachtsman and vice president of General Electric, who helped form a team of scientists, engineers, and industrialists.

In 1939, Putnam's team contracted with the Central Vermont Public Service Company, the first American electric utility to co-generate with an alternating-current windmill. With war approaching, designs were rushed into production before materials became scarce. A 2,000-foot peak in the Green Mountains of central Vermont was chosen: Known as Grandpa's Knob, this obscure site twelve miles from the town of Rutland became host to the largest wind turbine of the mid-20th century. Its 175-foot steel blades topped a 110-foot tower. This turbine rated an astonishing 1.25 megawatts in winds of about 30 mph, according to the U.S. Department of Energy. On October 19, 1941, the Smith-Putnam wind turbine delivered electricity to the power company for the first time.

The project marked an important milestone in wind-generated power. Unfortunately, during normal operation on an early spring day in 1945, one of the mill's giant blades few off, ending another great attempt to harness the wind. Rushed into completion before proper stress and wind tests could be completed, the massive turbine had failed, though much was learned.

In 1942, the Danish engineering company F.L. Smidth built two- and three-blade wind turbines, which played an important role in Denmark's electric program. But fuel prices fell again after the war, as did interest in wind turbines. This pattern of the price of fossil fuel prices see-sawing with the use of wind energy has continued. With the oil embargoes of the 1970s, came a sudden rush of money for research and development of wind power.

In the early 1980s, California built a large wind farm in Palm Springs. More than one-thousand wind turbines—half of them coming from Denmark—were installed. But by mid-decade, California withdrew its support, and, according to the Danish Wind Industry Association, the U.S. market for wind energy disappeared virtually overnight. Meanwhile, Europe continued to develop the field, with Germany installing the world's largest wind power farm and becoming the world's main market for wind turbines.

One small, tenacious New England town held its own during an apparently dismal time for wind energy in America. In 1979, rural Princeton in central Massachusetts decided to look for an alternative to burgeoning nuclear power. Unlike other towns in the area, Princeton declined to use electricity from the controversial Seabrook nuclear power plant in New Hampshire. The citizens of Princeton petitioned the town to form their own power department and to explore hydro, solar, and wind sources of energy. The town went on to purchase a good percentage of its power from water resources—and to build its own wind farm. The Princeton Municipal Light Department now lays claim to having America's longest continuously operating municipally owned wind energy facility.

Delores Lyons, the Chair of Princeton's Board of Selectmen, recalled the excitement of the early years: "Given the choice between Seabrook and windmills, it was a no-brainer." The town united in a loud message that it would take responsibility for its energy usage and oppose pollution. "No Nukes" bumper stickers appeared everywhere.

The wind farm on Wachusett Mountain, Princeton, MA, built in 1984. Photo by the author.

In early 1981, Princeton chose sixteen acres on the side of Wachusett Mountain as a possible site for a wind farm. The location was monitored for two years for wind velocity, direction, and consistency, as well as wildlife populations. In January of 1984, Princeton voted to acquire the site and, remarkably, in September of the same year a wind farm was up and running. Eight Enertech 100-foot towers were installed, with 22-foot-long blades, producing a total of 320 kilowatts of power. Original plans had called for 164-foot towers capable of producing 500 kilowatts of electricity, about 10 percent of Princeton's power needs; but due to the high cost, the town had to choose the smaller turbines which contributed no more than 2 percent of the town's needs. Nevertheless, for nearly twenty years the wind farm has stood in testimony to the will of a small group of dedicated people. The eight windmills are nearing the end of their useful life, and four of them are currently crippled by age and/or lightning strikes.

On a bright day in mid-March, I went to see the Princeton wind farm for myself.

Set in a quintessential New England countryside, only two hours from Boston, Wachusett Mountain is one of the busiest mountains in North America. A quarter of a million people visit its summit each year, ten thousand a day during the fall foliage season. In 1842, Henry David Thoreau wrote of finding "blueberry bushes, raspberries, gooseberries, strawberries, moss, and a fine, wiry grass" on the summit. He read Virgil and Wordsworth in his tent that night and imagined the mountain might someday be a Parnassus haunted by the Muses.

I met John Allard at the foot of Stage Coach Trail, just off of Westminster Road, on the side of Wachusett Mountain. Allard, the General Foreman of the Princeton Municipal Light Department, is known as "Little John," for his great size. He generously drew a map to guide me up the snowy trail, and allowed me access to the eight windmills and the control shack. This is in stark contrast to the heavy security necessary at nuclear power plants like Seabrook: the armed guards, the need to bombproof reactors, the radiation monitors that visitors must wear.

The Princeton, MA, wind farm from Stage Coach Trail, 2003. Photo by the author.

As I headed up Stage Coach Trail, I could hear the spring call of chickadees. About 100 yards from the top of the slope, I saw the blades of windmill number eight. Suddenly, it caught the wind and began to spin against the bright blue sky. I could hear the hum of the turbine that was soon muffled by the sound of the trees as a breeze picked up. As I drew close, windmills seven, five, four, and three came into view. Seven and three were spinning as well, until the wind died. With the blades still, the windmills seemed, if anything, more alive. Each turned slightly in the breeze, back and forth, feeling out the wind, like seagulls turning their heads, waiting for the moment to launch into the air.

When the wind picked up to about twelve miles per hour, first one, then another of the mills gave a groan and began to make a wobbling hum. Coming upon such a scene from the lower trail, I was reminded of my approach to the famous Old Mill of Nantucket. There, I had been awed by the turning sails of the 18th century windmill, and I was now equally moved by these wind turbines. The wind farm somehow retained some of the magic of the old wooden windmills. The old mills creaked like sailing ships; the modern windmills combine the stark beauty of kinetic sculpture with the homely, even soothing sound of a clothes drier.

Jonathan Fitch, the General Manager of the Princeton Municipal Light Department, has spearheaded a proposal for the next phase of the wind farm. He told me, "We made the hard choice to invest in wind power 18–20 years ago when it was still expensive and experimental. The technology change has produced more reliable, larger turbines that can produce renewable energy cost effectively." These advances in design and efficiency since 1984 have been such that the existing eight towers can be replaced with two larger windmills that would produce about 40 percent of the town's energy needs, nearly ten times the amount of power of the present wind farm. "Many of the issues we see in the news today," Fitch continued, "somehow relate back to our demand for oil. We need to remove that demand for oil from the equation before we can solve the other issues we face."

As proposed, each tower would be 230 feet in height, with a blade length of 118 feet, producing 1,300 – 1,500 kilowatts each. The existing towers produce only 40 kilowatts each. Wind velocity would be significantly greater atop the 230-foot towers, and the larger rotors would allow for fewer revolutions per minute, a gentle 17 – 20 rpm.

The Massachusetts Audubon Society, while fully supportive of renewable energy, has come out neither for nor against the project. Wachusett Mountain is enormously scenic. Most of the mountain lies within a protected state park, and much of the rest is either Audubon Society land or used by skiers. Some consider the windmill farm urban, industrial encroachment.

In 2002, as Princeton debated whether to go ahead with Jonathan Fitch's proposed upgrade of the facility, several voices were raised in opposition. Owners of a local inn, for example, expressed concern that wedding couples may not want to visit if the new, taller towers were visible from lawns where wedding photos are taken. Others raised the issue of avian deaths, contending that windmills slice birds like blenders. (Jonathan Fitch countered the "Cuisinart theory" by noting that bird deaths from cats, motor vehicles, and flights into glass windows number in the millions. The existing Princeton wind farm, he notes, has been responsible for fewer than three bird deaths per year per tower.)

On February 11, 2003, the town of Princeton considered a ballot measure for the upgrade of their wind farm. Proponents of the project held their breath as the vote was taken. Despite having no clear support from the Massachusetts Audubon Society, and some local opposition, Princeton voted overwhelmingly in favor of replacing the outdated windmills with two powerful new ones.

The innovative town of Princeton has a friendly rival in the eastern Massachusetts town of Hull. Located on the edge of Massachusetts Bay, it is another example of a group of people banding together to establish a municipally owned wind-power facility. In 1984, Hull set up an experimental mini-windmill. Fourteen years later, local residents formed the Citizens for Alternative Renewable Energy and successfully petitioned the Hull Municipal Light Department to erect a state-of-the-art wind turbine.

In April, 2001, Vestas (a Danish company) won the bid for a single, 150-foot-tall wind turbine rated at 660 kilowatts. On December 27, 2001, the Hull Light

Company's operations manager, John MacLeod, threw the switch and put Hull's windmill online. The next month, the Hull Light Department suspended billing for the town's streetlights, floodlights, and traffic lights. In its first two months of operation, the wind turbine banked over $20,000 worth of excess kilowatts. Within the first year, the facility won four awards from renewable energy associations.

The success of the Hull installation inspired other New England towns, such as Quincy and Barnstable in Massachusetts, to consider similar projects. Hull didn't stand still, however. On October 17, 2002, meetings were begun to consider the potential doubling of the output of the town's facility with the addition of another turbine.

In June of 1997, Searsburg, Vermont, a small, rural New England town, became the site of the largest wind power plant east of the Mississippi. Today, it is still considered the largest commercial wind farm in the northeast. Owned by the Green Mountain Power Company of South Burlington, Vermont, it has been producing enough wind energy to supply 2,000 homes.

From 1990 to 1994, the Green Mountain Power Company studied two test turbines it had built on Little Equinox Mountain in Manchester, Vermont. With data gathered from the site, along with input from public surveys, the company went on to identify over 400 Vermont hills and mountains over 1,500 feet in elevation. From among them, the company planned to choose the most suitable site for its new windmill farm. Those with environmental and land use conflicts were eliminated, as were those too remote from existing roads and transmission lines. Because of the possibility of falling ice, sites close to major hiking trails were dropped, as well as those near ski areas.

The next phase involved the installation of wind measurement equipment on each of eight proposed sites. Biologists, botanists, and civil engineers conducted studies that helped narrow the choice to four sites. In 1993, the Searsburg area emerged as the most viable. In September of 1996, the first wind turbine was lifted into place and in only eight months a wind farm was completed.

Eleven 550-kilowatt wind turbines sit along a ridge of Mount Waldo, about 17 miles east of Bennington in south central Vermont. Zond Systems, Inc., of California designed and manufactured the windmills and presently maintains them. Each tower is 132 feet high, and the blades, 66 feet in length, revolve at 29 revolutions per minute. Unique to northern wind farms, the blades are painted black to help diminish the formation of ice.

Before the installation, bird and bear habitats in the area were studied. It was found that few hawks or songbirds migrated though the immediate countryside. (In any case, most bird migration takes place well above the height of the turbines.) Unlike the earlier lattice towers, the newer single-column towers have no areas where birds might be attracted to perch.

When I spoke to Martha Staskus about touring the wind farm, she described the company's concern with the local bear population. Staskus is a representative of the

Vermont Environmental Research Association and a consultant to Green Mountain Power. She told me of two populations of bears located in the forests to the east and west of the wind farm. No public access is allowed to the site during the deep winter snows, or when bear cubs emerge in the spring. Typically, one public tour is conducted in May or early June, then others are offered periodically throughout the summer and early fall between July 15 and October 15.

I saw the Searsburg Wind Power Facility at the end of winter in mud season. Turning off Route 9 in Southern Vermont, I drove about three miles south on Route 8. To the left, I could see several of the towers peeking through the bare trees. The welcome sign for the facility is on the left, at the beginning of Sleepy Hollow Road. Just above this sign, five of the eleven wind turbines can be seen. To the left, I could see only the blades of two others above the tree line, turning lazily, much like the blades of the old windmills that peek above the hedgerows of the Netherlands.

I went into the sub-station to meet the aptly surnamed Art Miller, supervisor of the facility. I knew that Art worked alone, caring for the windmills, surrounded by dense green mountains. At first, I likened him to the old millers who sometimes worked long hours in solitude. But the old windmills were active community centers as well, where farmers came and went, exchanging village news along with grain and flour. Art Miller, I imagined, lived more like the old lighthouse keepers, and this was not far from the truth.

Instead of block-and-tackle, grain hoppers, and four-ton millstones, Art works surrounded by computer systems that monitor every aspect of his windmills. Art began his professional life working for a petroleum contractor, building and repairing gas stations in Los Angeles. In the summer of 1997, he relocated to the Southern California desert and saw his first wind turbines. The sight of the clean lines of the turbines, and the thought of the endless energy they could provide, changed his career plans.

Now, happily working with wind rather than petroleum, he described the technical advances made in the last few years. He began by comparing wind turbines to automobile engines: "If you're a mechanic and you work on cars—you're not happy; they're very hard to work on. The electro-mechanical parts of the wind turbines are ergonomically designed. They're *made* to be worked on."

Miller said the turbines in Searsburg have been made "relatively" obsolete by the new, larger turbines that produce nearly triple the power. The General Electric company's new 1.3-megawatt turbines, he explained, have variable speed generators and variable pitch blades that vary their pitch according to wind conditions. The new generators produce electricity within a much broader wind spectrum. Both blades and generators make it possible for the new turbines to generate power at much lower *and* much higher wind speeds. Preliminary plans are already being made to produce a proposed twenty to thirty megawatts of power at Searsburg from about fifteen new 1,500-kilowatt wind turbines. Art also mentioned that wind turbine manufacturers have realized in the past eight years that "fashion is as important as function—the new machines are more attractive."

Art confessed that "the hardest part of my job is climbing the towers." This is done from within each tower, with a fully trained and certified partner who must be called in from the central office. Thus, Art Miller, like the old windmillers, still has some dangerous climbing to do.

Art handed me a hard hat and offered to take me to see the wind turbines up close. He had to do a minor repair on turbine number six, and I was eager to take some photos. At the end of the muddy access road, a line of towers appeared, ringed magnificently by Vermont's southern hills. Rising half again as high as the turbines in Princeton, Massachusetts, these have a galactic presence; though fully grounded on the Earth they wouldn't look out of place as part of a NASA expedition to a distant planet. Unlike the older Princeton turbines, these were eerily quiet.

Art and I spent a good part of the afternoon talking about everything that seemed to be under the sun that day. He told me about his life, his family, and his work. When he talked of his hopes for wind power—its potential to link with hydro, solar, and fuel cells—he spoke of the Searsburg windmills as if they were living things. "They're restless if they're idle," he told me. "Static electricity builds up and this fouls up the communication components. If the wind picks up to 29 or 30 miles per hour, they're tickled pink. They just love it."

Art Miller in the doorway of a wind turbine tower at the Searsburg Wind Facility, VT. Photo by the author.

Before I could thank Art Miller for his generosity, he thanked me for taking so much time with him. Miller said, "You know, I go a little stir crazy up here." I snapped a photo of Art in the doorway of one of the round, white towers. He looked less like a windmiller than a 21st century lighthouse keeper.

By the end of the 20th century, engineers began sending windmills to sea when they discovered that 20 per cent more energy could be produced in water than on land. In 1991, the Vindeby wind farm was built in the Baltic Sea off the coast of Denmark. According to the Danish Wind Industry Association, "The park has been performing flawlessly," and another farm was built off the coast in the Kattegat Sea in 1995.

None of the above developments came without protest. When the Kattegat Sea installation was announced in 1992, people on the eastern part of Jutland voiced concerns about its effects on marine ecology and the property values of seaside cottages, as well as the potential annoyance of light reflection and noise. In 2002, New England became intimately aware of these same issues. On Block Island, and particularly in Cape Cod's Nantucket Sound, proposed wind farms pitted local residents, summer cottage owners, environmentalists, politicians, and corporations against one another in various surprising combinations.

Located off the coast of Rhode Island, Block Island provides a good example of New England's ambivalence toward wind power. In 1979, the U.S. Department of Energy set up an experimental wind power station on the island to test equipment and conditions. The National Aeronautics and Space Administration developed the turbine and blades used there. According to the *Block Island Times*, "H.G. Wells would surely have been inspired by this gargantuan erector set constructed on the grounds of the Block Island Power Company. The blade was 125 feet across – 20 feet longer, for example, than the Harborside Inn on Water Street." [2]

One of only four 200-kilowatt windmills in the United States, the Block Island mill provided 18 percent of the island's electricity. The experiment was funded for only two years, after which the facility languished. The U.S. government sold the $2.3 million complex to the local power company in 1984 for $1. The Block Island Power Company then sold the windmill for scrap for $2,000.

Some Block Island residents set up private windmills in the 1980s and 1990s. But when a survey was taken of local sentiment in early 2001, the results proved far from encouraging. 70 percent of the Block Island Residents Association said that "preserving the current visual and audio environment of the island" was more important than reducing power costs or fuel use. Only 11 percent supported "making progress toward reducing the cost of power," and only 10 percent supported "making progress toward decreasing the use of fossil fuel on Block Island." 70 percent said they would mind having one of the new windmills near their property, and 73 percent agreed when asked, "Do you think the presence of a new windmill next to your property would decrease the value of your property?" [3]

A major turning point in the evolution of wind power took place in New England on November 15, 2001. On that day, the first formal application in United States history was made for an offshore wind farm. An entrepreneurial energy company called Cape Wind Associates proposed placing 170 turbine windmills in Horseshoe Shoals, the windy shallows in Nantucket Sound. This proposed wind park, when completed, would spread over an area of about twenty-five square miles.

Cape Wind Associates estimates that the facility will produce about two-thirds as much power as is generated by the Pilgrim Nuclear Power Plant in Plymouth,

Massachusetts. In other words, at its peak hourly output, the wind farm would provide all of the energy used by Cape Cod and the islands of Nantucket and Martha's Vineyard.

This $700 million project is now about halfway through a rigorous federal permit process and, if built, is expected to be completed in 2005. Deborah Donovan of the Union of Concerned Scientists in Boston is quoted in the *Los Angeles Times* as saying, "I hate to make it sound like this is going to save the world, but it really does set the stage for what is going to happen elsewhere." [4] Since Cape Wind Associates made its proposal, ten other wind power projects are now under consideration in the New England region alone.

There is, however, fierce opposition to the Cape Wind proposal. The Alliance to Save Nantucket Sound and the Nantucket Chamber of Commerce, along with other organizations, have raised multiple concerns. One common consideration is whether ocean views will be obstructed, and how this would affect real estate values. Cape Wind President James Gordon counters that the wind farm would be situated six miles from Hyannis and thirteen miles from Nantucket. From these distances, he claims, the towers "would look like tiny, tiny masts on the horizon—but only on a clear day." [5]

Nantucket Sound is, of course, a pristine cruising ground for pleasure boats and yachts. Some argue that a more than 25-square-mile area filled with 270-foot-tall windmills would be a navigational nightmare. Both Hyannis and Nantucket have airports, and air traffic controllers fear that the windmills may come afoul of small aircraft. Scallop fishermen, already under pressure from declining stocks, worry for the future of their scallop beds.

Senator Ted Kennedy, whose home overlooks Nantucket Sound, opposes the project, One of Martha's Vineyard's most famous residents, former CBS news anchor Walter Cronkite, has struggled with the issue. Initially publicly opposed to it, he has since decided to wait for the results of various governmental studies.

Determining the future of wind energy in New England is no easy task. Polling the public has only served to embarrass the pollsters, whose vastly different figures serve only to confuse people. In November of 2002, Cape Wind Associates released their survey of residents both on and off Cape Cod. Not surprisingly, the company found that 55 percent of on-Cape residents supported the wind farm, and 35 percent did not. Off-Cape, there was even more support: 64 percent for, 22 percent against.

The release of Cape Wind's figures prompted the opposing Alliance to Protect Nantucket Sound to release a survey they commissioned through the private polling firm of Harrison and Goldberg of Cambridge. This firm found that on Cape Cod, 35 percent supported the wind farm and 58 percent were opposed. Off-Cape, 28 percent were in support, and 66 percent were opposed.

Two sides of the issue, two surveys, and two diametrically opposite results.

The director of the world famous Woods Hole Research Center on the Cape, biologist George M. Woodwell, has weighed in on the side of wind farms: "I think one has to look at the issue, and the issue is how to get rid of fossil fuels as fast as possible. We have an Earth that is warming very rapidly." [6]

Environmentalists in New England and across the country see the Cape Wind's proposal for Nantucket Sound as critical to the future of wind energy in the U.S. Greg Watson of the Massachusetts Technology Collaborative goes one step further: He feels the entire world will be watching how state and federal regulators evaluate the Nantucket Sound project. Watson has said, "If it's done poorly and [fulfills] people's fears, it could set renewable [energy] back decades, or perhaps so far that they couldn't recover." [7]

Opposition to one particular wind proposal doesn't necessarily mean opposition to all others, however. The Cape Cod town of Barnstable has come out against the Nantucket Sound proposal, but has a proposal of its own: Four wind towers, each 200 feet tall, would provide electricity for the town's sewage treatment plant and two nearby Department of Public Works buildings. DPW Director Thomas Mullen told a Town Council meeting, "We're talking about windmills at the wastewater treatment plant, we're talking about turbine engines that will produce heat and electricity, we're talking about all kinds of energy saving devices." [8] At present, the town pays $300,000 per year for the electricity needed to run its treatment plant and DPW buildings.

On Martha's Vineyard, I spoke to environmental researcher and writer William Marks. He sees wind power as a necessary component for the survival of humanity. "There is only so much living energy in our biosphere," he states. Marks, however, considers wind power a transitional source of power, one which will be eclipsed in the future by hydrogen fuel cells.

One New England organization has put itself forward as a possible mediator among opposing factions at this critical time: The Massachusetts Technology Cooperative was established by the State Legislature in 1982 (as the Massachusetts Microelectronics Center). Today, the cooperative hopes to become a trusted, neutral source for information on renewable energy. For the Nantucket Sound project, it hopes to bring together the solar wind industry, the Massachusetts Audubon Society, local yachtsmen, the League of Women Voters, the Cape Cod Commercial Hook Fisherman's Association, and more than a dozen other groups. Moreover, the cooperative is helping place wind power within the complete range of renewable energy sources, including wave and thermal energy from the sea, methane production, fuel cells, and solar voltaics.

The Nantucket Sound controversy has forced environmentalists to awkwardly protest an energy project that would normally be supported—were it placed elsewhere. Given the vast length of the eastern coastline, why, then, was Nantucket Sound chosen? Dozens of European offshore wind farms have been built in more isolated waters with the backing of governmental funds. Without federal funding, Cape Wind Associates needed to find a location both close to the power grid and naturally protected from high seas. From an economic standpoint, Nantucket Sound seemed ideal. Opponents hope to stall the project until the U.S. can adopt the European pattern of government and private partnership. A strong federal commitment would make wind farms feasible in the isolated plains of the American West or the remote offshore areas of New England.

Wind energy is, according to the U.S. Department of Energy, the fastest growing source of energy worldwide. The Department, not generally known as a great supporter of alternative energy, has said that "the lessons learned from more than a decade of operating wind power plants...have made wind-generated electricity very close in cost to the power from conventional utility generation in some locations." [9] Since that statement was made, costs have dropped further. On June 20, 2003, John Dunlop, a regional manager of the American Wind Association, made a remarkable assertion on National Public Radio's "Talk of the Nation": Wind, he said, can supply energy to power companies at less cost than any other source of power. In other words, wind power has become the cheapest form of energy available.

North and South Dakota, according to Dunlop, could produce 80 percent of all energy needs for the U.S. The major impediment to this extraordinary scenario is the lack of infrastructure. Transmission lines from remote places like the Dakotas don't yet exist to transport wind energy to major power grids.

In the 21st century, we're still capable of being awed by the rare sight of an historic windmill carrying sixty feet of sail over slowly turning blades. Children raised on modern technology still marvel at the centuries old sound of creaking and straining wood, and the indescribable aroma of corn being ground between massive granite millstones.

The windmill is still very much in evidence in our society. We often see them on lawns or in gardens, but more often windmills reside in the realms of our imaginations, in our language, and in the enduring imagery of poetry, literature, and music. We are currently in the midst of a great resurgence of the windmill, one potentially as important as its ancient adaptation for the production of bread.

In early 2003, Denmark completed the world's largest offshore wind farm: Eighty wind turbines have been erected at Horns Rev in the North Sea. Located from nine to twelve miles off the western shore of Denmark, they are just barely visible from shore. This 160-megawatt wind farm provides 2 percent of all of Denmark's power needs. In Great Britain alone, eighteen offshore wind farms are under development. I recently drove past six inland wind farms in the space of a single afternoon in southwest England.

In the U.S., wind farms continue to expand. In New England, as noted, more than ten wind farms—both on- and offshore—are in development. In the state of Washington, Desert Claim Wind Power, a subsidiary of enXco, Inc., has proposed a 180-megawatt wind farm; consisting of 120 wind turbines, it will lie within a major cross-state electric transmission corridor, linking the hydroelectric power produced by the Columbia River with the large power consumer market of western Washington. And, ironically, General Electric, which in 1892 owned the wind power company begun by pioneer Charles Brush, has recently bought Enron Wind and has re-entered the field of wind energy.

On March 25, 2003, Greg Abel, president of MidAmerican Energy Holdings Co., made a startling declaration in Des Moines, Iowa: "Today we're here to announce that the largest wind facility to be constructed in the world will be built in Iowa." [10] From 180 to 200 wind turbines will be erected on 200 acres of northern Iowa farm fields by the year 2006. The 310 megawatts of electricity generated will be enough to power 85,000 homes.

Nearly 2,000 years ago windmills were first documented in Persia. Since then, the wind, whether sent by the gods or produced by colliding air masses, continues to be captured by windmills. The intense romance of Nantucket's Old Mill can't yet be matched in the imagination by the modern, sleek propellers of the kind proposed for Nantucket Sound. Perhaps the windmills of the 21st century are just waiting for a Cervantes to inspire our vision. An epic story of a modern-day Don Quixote, sailing into Nantucket Sound to do battle with the windmills he mistakes for giants, may help us re-imagine the options of our contemporary energy dilemma.

We can do worse than to remember the words of Miguel de Cervantes:

> *To imagine that things in this life are always to remain as they are is to indulge in idle dream. It would appear, rather, that everything moves in circles.* ●

"How the Mill Turns and the Cabbages Grow," from *Mr. Wind and Madam Rain*, a book of Breton folk stories by Paul de Musset, NY: Harper & Brothers, n.d.

Appendix One: *Day Trips Along the Windmill Trail*

One can take in some of Southern New England's most spectacular coastal scenery and hidden villages by linking together windmills. The Windmill Trail I've set out here takes us on day trips to rare and awe-inspiring authentic windmills and many irresistible and imaginative reproductions. Most public windmills are run by volunteers and have limited hours, which may change unexpectedly. Call the contacts noted for each mill ahead of time, if possible. Reproduction windmills, unless stated, are privately owned and should be enjoyed without disturbing the owners' privacy or trespassing on private property. Most owners are happy to see visitors—from a distance. For complete, detailed information on each of the mills, refer to previous chapters.

The great stretches of the southern New England coastline—where winds are high and rivers are few—have proven ideal locations for windmills. Cape Cod and the Islands, and the romantic coast of Rhode Island, are places to be savored slowly. Following this Windmill Trail is an ideal way to experience a land, an ocean, and a way of life that are unique in the world.

Day One

CAPE COD

BOURNE – CATAUMET – SANDWICH

Joseph Jefferson's windmill at the Aptucxet Trading Post Museum, Bourne, MA. Photo by the author.

BOURNE

A good way to begin any tour of Cape Cod is at the often-overlooked Aptucxet Trading Post Museum in Bourne. Here, the Pilgrims established their trading post in 1626 to trade with both the Dutch colonists of New Amsterdam and the Native American population. The museum's replica trading post, windmill, and herb and wildflower gardens front the dramatic Cape Cod Canal.

The Windmill

At the entrance to the grounds of the **Aptucxet Trading Post Museum** is a Dutch-style reproduction windmill.

Built by actor-artist Joseph Jefferson in the late 19th century at "Crow's Nest," his home in Buzzards Bay, it now serves as a gift shop. Jefferson, well-known for his portrayal of Rip Van Winkle, used the windmill as his art studio.

Location

Aptucxet Trading Post Museum
24 Aptucxet Road
Bourne

Directions

Cross the Bourne Bridge to Cape Cod. At the Bourne rotary just after the bridge, take the exit for Trowbridge Road. Follow Trowbridge Road approximately one mile to the stop sign at the intersection of County Road. Continue straight onto Shore Road for approximately one mile. Watch for a sign on the right for the Aptucxet Trading Post. Turn right, go under the railroad bridge, and take an immediate right onto Aptucxet Road. The windmill at the entrance of the museum is visible fifty yards on the left.

Hours and Entrance Fees

Open from Mid-April to Mid-October, Tuesday through Saturday from 10 am to 5:00 pm and Sundays from 2:00 pm to 5:00 pm. Also open Mondays in July and August, closed Memorial Day, Labor Day, and Columbus Day.
Nominal entrance fee to museum. Special rates available for senior citizens, students, group tours. Children under 6 admitted free.

Contact for Information

The Bourne Historical Society
Jonathan Bourne Historical Center
30 Keene Street
Bourne, MA 02532
Mail: Box 95, Bourne, MA 02532-0795
Tel.: 508-759-8167
Web Site: www.bournehistoricalsoc.org/wgs.htm

Other Features

Reproduction saltworks, Aptucxet Trading Post Museum, Bourne, MA. Photo by the author.

Also on the grounds you will find the only reproduction saltworks on the Cape. Hundreds of these small, rudimentary windmills once lined the shores of Cape Cod to pump seawater into wooden vats to be evaporated into salt. President Grover Cleveland owned a summer home in Bourne, and in 1892, the Old Colony Railroad built a station for his exclusive use. This Victorian railroad station was later moved here to the museum.

The Red Brook Mill, Shore Road, Cataumet, MA. Photo by the author.

BOURNE – CATAUMET VILLAGE

Cataumet has three remarkable authentic windmills—two with romantic literary traditions, and one a transplanted Rhode Island giant.

The Windmills

The **Red Brook Mill** on Shore Road was built in Bristol, Rhode Island, in 1797, and moved to Fairhaven, Massachusetts, in 1821. Located just across from the whaling port of New Bedford, the mill was used for sharpening harpoons. The mill moved once again when the Gammons family of Bridgewater purchased it in 1905. The next year Ferdinand Gammons had cosmetic restoration done and moved it to his summer home on Shore Road, Cataumet, where it currently stands.

In 1900, successful Boston insurance man John J. E. **Rothery** bought a windmill that had been built in Chatham around 1730, near the present location of the Chatham Bars Inn, and moved it to his summer estate in Cataumet. Rothery very shortly had another windmill moved to his estate, and both mills can be seen in old photos before trees and other houses obscured the view. When his daughter, Agnes, grew up, she used the estate as the setting for her fictional works like *The House by the Windmill*, published in 1923.
The first mill is on Red Brook Pond Drive and is used as a living space and artist's studio; and the second, on Old County Road, is used as guest quarters and sometimes a guitar maker's studio.

Early postcard of the **Rothery Windmill** on County Road, Cataumet, MA.

Locations

The Red Brook Mill
1094 Shore Road
Cataumet Village, Bourne

The Rothery Windmills
9 Red Brook Pond Drive
and
1090 County Road
Cataumet Village, Bourne

Directions

From the windmill at the Aptucxet Trading Post, exit right onto Aptucxet Road. At the first intersection, turn left and go under the railroad bridge. Take an immediate right on Shore Road. Continue along scenic Shore Road to Cataumet Village. At approximately five miles, Red Brook Pond will be on the left. On the right, at 1094 Shore Road, will be the Red Brook Windmill. Continue on Shore Road one quarter of a mile to County Road and turn left. Take the second left onto Old Mill Lane, then another left onto Red Brook Pond Drive. The Rothery Windmill will be immediately on the right, at 9 Red Brook Pond Drive. Return to County Road and turn left. Watch for the other Rothery Windmill on the left, at 1090 County Road.

Rothery Windmill, County Road, Cataumet, MA. Photo by the author.

Access

All are privately owned. View from road, do not trespass.

Further Information

The Bourne Historical Society
Jonathan Bourne Historical Center
30 Keene Street
Bourne, MA 02532
Tel.: 508-759-8167
Web Site: www.bournehistoricalsoc.org

From Pocasset to Cataumet, published by the Bourne Historical Society.
Agnes Rothery, *The House by the Windmill*, NY: Little, Brown, 1923.

SANDWICH

Heritage Plantation, in the splendid town of Sandwich, is the home of one of Cape Cod's most historic remaining windmills.

The Windmill

The **Old East Mill** was moved from Orleans to the Heritage Plantation Museum in Sandwich in 1968. It had been, up to that time, the last remaining windmill in Orleans. On the way to Sandwich, notice the large reproduction windmill of the Christmas Tree Shop at the foot of the Sagamore Bridge. It was built in the early 1980s as an operating windmill, but because of alignment problems, it never worked.

Location

Heritage Plantation of Sandwich, Inc.
Grove and Pine Streets
Sandwich

Directions

From the Rothery Windmill in Cataumet, continue north on County Road. Turn right onto Long Hill Road. Next, turn left and follow Route 28A to the entrance to Route 28 at Pocasset. Take Route 28 to Bourne, and from the Bourne rotary take the first exit on the right, onto Route 6. Follow Route 6 along the Cape Cod Canal to Sagamore, where 6 becomes Route 6A. From Route 6A take Route 130 to Pine Street. Follow signs to Heritage Plantation at Pine and Grove Streets.

Hours and Entrance Fees

Open mid-May until late-October, seven days a week, 10:00 am to 5:00 pm.
Adults $9.00, Seniors $8.00, Children 6 – 8 $4.50, Five and under free.
The interior of the windmill is not open to the public.

Contact Information

Recorded Information Line: 508-888-1222
Tel.: 508-888-3300
Fax: 508-833-2917
Web Site: www.heritageplantaion.org
E-mail: heritage@heritageplantation.org

Other Features

In addition to the Old East Windmill, Heritage Plantation has a dazzling seventy-six acres of Dexter rhododendrons, daylilies, hostas, and heathers; a reproduction Shaker round barn that houses an antique car collection; a 1912 carousel; and a military museum.

Day Two
CAPE COD
OSTERVILLE – HYANNIS PORT – SOUTH YARMOUTH
WEST HARWICH

Day Two begins in the out-of-the-way jewel of Osterville. Overlooked by most guidebooks, it's one of the lovelier villages of Barnstable and contains two noteworthy reproduction windmills. The tour continues to the remarkable Kennedy family neighborhood of Hyannis Port. After a visit to the historic Judah Baker Windmill in South Yarmouth, the day will end, hopefully, with a stunning sunset view of the Windmill House at Old Mill Point, West Harwich.

Oyster Harbors Windmill, Osterville, MA. Photo by the author.

OSTERVILLE
The Windmills
In 1925, **Oyster Harbors** island was chosen by Forrest W. Norris and a group of investors as one of the most desirable spots on the Cape, and an exclusive summer colony soon developed. Nearly two million dollars were put into facilities that included a golf course, tennis courts, stables, roads, dredged harbors, and a large clubhouse. Approaching the island over a causeway, one can still see a gatehouse on one side of the island's entrance, and a reproduction windmill on the other.

There is a little gem of a windmill on a narrow road called Cockachoiset Lane, just by the Osterville Yacht Club and Crosby's Boat Yard. Attached to a lovely shingled house is **Le Petit Moulin**, a reproduction windmill built for Boston architect and interior designer Richard Fitzgerald in the 1950s. This is in all probability one of Ivan Kendrick's mills—Kendrick was the premiere Cape Cod reproduction windmill builder of his era.

Locations
Oyster Harbors Windmill
Bridge Street
Osterville

Le Petit Moulin
12 Cockachoiset Lane
Osterville

Access
Both windmills are privately owned. View from road.

Directions
From Falmouth take Route 28 toward Hyannis, and turn right on County Road to Osterville. Alternately, from Hyannis take Route 28 toward Falmouth, and turn left on County Road, Osterville. Follow County Road through Osterville center to Main Street. Turn right at West Bay Road (the Osterville Public Library will be on the right) and follow signs to the Osterville Yacht Club and Cosby's Boat Yard. Turn left onto Bridge Street (at the boat yard) and then immediately right on Cockachoiset Lane. Le Petit Moulin will be on the right. Exiting Cockachoiset Lane, turn right on Bridge Street and follow it to the causeway and gate of Oyster Harbors, where you will find its white windmill.

Further Information
Osterville Historical Society
155 West Bay Road
Osterville, MA 02655
Tel.: 508-428-5861

Paul L. Chesbro, *Osterville*. Taunton, MA: William Sullwold Publishing, 1988.

Other Features
Leaving Osterville on Route 28 toward Falmouth, take a detour on the West Barnstable Road (Route 149) toward Marstons Mills, if time permits. In about two miles, at the intersection of Race Road, you will see Cape Cod Airport with its white reproduction windmill among the aircraft.

The Wright Windmill, Hyannis Port, MA. Photo by the author.

HYANNIS PORT

Hyannis and Hyannis Port may no longer have their great old windmills, but they do have some of New England's best, most eccentric reproductions. Hyannis Port is, of course, famous for the Kennedys. Their compound is within sight of several of the windmills.

The Windmills

The **Wright Windmill** and the house next to it on Hyannis Avenue were built around 1900 by George Wright of St. Louis. Among the more famous residents of the property was Boston Red Sox star Jim Piersall, whose eight children scrambled through the rambling house and its windmill.

Landmark House and its windmill tower were originally built in the late 19th century by George B. Holbrook of Springfield, Massachusetts. During Labor Day weekend of 1910, the shingled summer home was completely destroyed by fire. The only thing left standing was the windmill. An elegant new home was built, a sprawling, European affair with Doric columns and a pergola. Today, the tower looks like a graceful, pale yellow Dutch windmill without arms. Eight-sided, round-topped, with round portal windows under the eaves, and a gallery or "stage" around the bottom, it is used as an art studio by the modern impressionist painter, Samuel Barber.

The Landmark House and windmill tower, Hyannis Port, MA. Photo by the author.

Locations

Wright Windmill
60 Hyannis Avenue
Hyannis Port

Landmark House
10 Hyannis Avenue, corner of
Washington Ave.
Hyannis Port

Directions
From Osterville, return to Route 28 and turn right toward Hyannis. At Hyannis, turn right off Route 28, onto West Main Street. At Sea Street, turn right. This bends to the right and becomes Ocean Avenue. At 85 Ocean Avenue, on the left, notice the windmill attached to the house. It was designed by John Barnard in 1962 as additional bedrooms for the house. From Ocean Avenue, turn left on Hyannis Avenue. Immediately to the left will be a lane which leads to the Wright Windmill. Continue down Hyannis Avenue and directly ahead will appear Samuel Barber's windmill tower at Landmark House. Continue on Washington Avenue, past the public beach, to the neighborhood of the Kennedy Compound. On Wachusett Avenue, near the little red post office, is a decorative shingled windmill without arms. Continue on to 61 Dale Avenue, where a large reproduction windmill is located by the water.

Access
All of the windmills mentioned are privately owned and can be enjoyed only from the public roads, or from the water. Do not trespass onto private property.

Further Information
Web Site: www.hyannis.com

Paul Fairbanks Herrick, *Old Hyannis Port*. New Bedford: Reynolds-DeWolf, 1968.

Postcard of the **Judah Baker Windmill,** South Yarmouth, MA.

SOUTH YARMOUTH

When leaving Hyannis, notice the old railroad station on Main Street. To the right, you will also see Cape Cod Storyland and its reproduction of the Eastham Windmill. Along Route 28 in Yarmouth, notice the reproduction windmill on the left at a shop called Windmill Crafters. Built in the 1950s, it was used as a real estate office and has since moved twice before becoming part of this shop.

Yarmouth lost an important windmill when the Farris Mill was bought in 1935 and moved to the Henry Ford Museum in Dearborn, Michigan. Fortunately, in 1953 the town was able to acquire the Judah Baker Windmill, one of the Cape's finest and oldest remaining windmills.

The Windmill

The eight-sided **Judah Baker Windmill** was built in South Dennis in 1791. After moving at least four times, and undergoing three renovations in Yarmouth, it now overlooks the water at the foot of Willow Street at Bass River, South Yarmouth.

Location

The Judah Baker Windmill
Windmill Park
River Street
South Yarmouth

Directions

From Hyannis, continue on Route 28 through West Yarmouth to South Yarmouth. Just before the Bass River Bridge, turn right on Union Street. Take the first right onto Pleasant Street and continue for about one mile. Bear left onto River Street, and the windmill will appear by the water on the left.

Hours and Contact Information

Open twice weekly Memorial Day through Columbus Day. For information contact:
Yarmouth Town Offices
Building 1146, Route 28
South Yarmouth, MA 02664
Tel.: 508-398-2231, ext. 292

Further Information

Historical Society of Old Yarmouth
11 Strawberry Lane
P.O. Box 11
Yarmouth Port, MA 02675
Tel.: 508-362-3021
Web Site: www.hsoy.org

Postcard of **Windmill House,** West Harwich, MA.

WEST HARWICH

End the day in West Harwich at the mouth of the Herring River where you will see one of the most photographed windmills on the Cape.

The Windmill

The Windmill House at Old Mill Point (also known as Doble's Point), West Harwich, was built with the romantic photographer in mind. W.H. Doble developed a private enclave off Lower County Road in the 1920s which included reproductions of English and Dutch style homes, complete with authentic details like a sagging roof line. The Sugden family has owned the Windmill House since 1950.

Location

The Windmill House
Old Mill Point
5 West Strandway
West Harwich

Directions

From South Yarmouth, continue on Route 28 to West Harwich. At the center of West Harwich, turn right on Riverside Drive. As Riverside Drive approaches Nantucket Sound, turn left on Lower County Road. Continue to where Lower County Road crosses the Herring River. Turn off the road and park for a fine view of The Windmill House on the right at the mouth of the Herring River.

Access

Old Mill Point and The Windmill House are privately owned.
Please do not trespass.

Further Information

Harwich Historical Society
80 Parallel Street
Harwich, MA 02645
Tel.: 508-432-8089
Web Site: www.capecod.com/history
E-mail: harwichhistoricalsociety@capecod.com

Other Features
You can continue along Lower County Road to see a fine, detached reproduction windmill at 230 Lower County Road on the left. It is on private property.

Day Three
CAPE COD
BREWSTER – EASTHAM – ORLEANS – CHATHAM

This tour is packed with some of the most stunning of New England's oldest standing windmills. It heads towards Brewster along the historic Old Kings Highway (Route 6A), considered one of the ten most scenic roads in the country. It winds past farmsteads, cranberry bogs, sea captains' homes, and village greens. At one time the route was dotted with windmills and dozens of saltwork mills. (John Sears of Dennis was credited with inventing the solar evaporation method of producing salt in 1776, which used windmills to pump seawater into evaporation vats.) Vintage windmill garden ornaments can be found in the numerous antique shops along this route. In West Brewster, the Old Higgins Farm Windmill will appear on the left in a lovely field of wildflowers in Drummer Boy Park.

Old Higgins Farm Windmill, Brewster, MA. Photo by the author.

BREWSTER
The Windmill
Around 1795, the **Old Higgins Farm Windmill** was built in Brewster at the head of Ellis Landing Road. Little is known of the origins of this gray-shingled mill with an elegant, boat-shaped cap. We do, however, know why it had to be moved from its original location: The creaking and cracking of its gears and the whirling of its sails scared the heck out of passing horses.

Location
Old Higgins Farm Windmill
Drummer Boy Park
Main Street, Route 6A
Brewster

Directions

Travel east on Route 6A (the Old Kings Highway). After passing Dennis, watch for Drummer Boy Park on the left in West Brewster.

Hours and Entrance Fees

Hours vary. Contact the Brewster Historical Society for information.

Contact Information

The Brewster Historical Society Museum
3371 Main Street
Brewster, MA 02631
Tel.: 508-896-9521

Other Features

Also in Drummer Boy Park is the Harris-Black House. The plaque reads, "Harris-Black House, 1795. Possibly the last remaining primitive one room house on Cape Cod...."

Herring Run and its fish ladders are a spectacular sight during spring fish runs. It's next to the John Gorham gristmill at Satucket and Stony Brook Roads. The Gorham gristmill was built in 1873 and is still in operation.

Windmill Weekend, September, 2002, at the **Eastham Windmill,** Eastham, MA. Photo by the author.

EASTHAM

There is hardly a prettier town common in New England than the Eastham Common in the spring, when its enchanting windmill is rigged with sails and the daffodils are in bloom.

The Windmill

For generations, the origins of the **Eastham Windmill** were lost. It was clearly very old, having a conical cap in the older Flemish style, rather than the Kentish boat-shaped caps of other mills. It had been built by Thomas Paine, Eastham's famous millwright, and many logically assumed that he built it somewhere in his own town. However, recent opinion, supported by miller James Owens and historian Frederika Burrows, holds that Paine built the mill in Plimoth in the 1680s. Around 1770, it was dismantled for the first, but hardly the last, time and ferried across Massachusetts Bay to Truro on a log raft. Miller Seth Knowles brought it to Eastham in 1793. James Owens, the dean of Cape Cod windmill historians and the keeper of the Eastham Windmill, has stated, "It's hard to authenticate it but I do think it's the oldest mill in the country. If the 1680 date can be authenticated it definitely is the oldest mill in the United States."

Location
Eastham Common
Route 6 and Samoset Road
Eastham

Directions
From Route 6A, continue through Orleans (we will return to Young's Windmill in Orleans later) to the Orleans rotary. Enter the rotary and take the first exit on the right to Route 6, Eastham. The windmill is easily visible on Route 6, on the left at the corner of Samoset Road in the center of town.

Hours and Entrance Fees
Open generally during the summer from 10:00 a.m. to 5:00 p.m., and during Windmill Weekend in September. Check with Eastham Town Hall for specific times and fees.

Contact Information
Eastham Town Hall
2500 Route 6
Eastham, MA 02642
Tel.: 508-240-5900

Other Features
Poit's Family Restaurant and Mini-Golf in North Eastham has a good example of a plywood miniature golf windmill. Set among a wishing well, bridge, and large figures of Porky Pig and Uncle Wiggly, this simple mill has a small motor that moves its arms.

The Jonathan Young Windmill, with van from Windmill Liquors in foreground, Orleans, MA. Photo by the author.

ORLEANS

In October 1849, writer/naturalist Henry David Thoreau stayed at Higgins Tavern on the County Road in Orleans. In his book, *Cape Cod*, he wrote of seeing two young Italian organ grinders, and of the "saltworks scattered all along the shore, with their long rows of vats resting on piles driven into the marsh, their low turtlelike roofs, and their...windmills, novel and interesting objects to an inlander." Of traditional smock windmills, Thoreau wrote, "They looked loose and slightly locomotive, like huge wounded birds, trailing a wing or a leg, and reminded one of the pictures of the Netherlands."

The Windmill

The **Jonathan Young Windmill**, overlooking Town Cove, is one of the Cape's earliest and best preserved windmills. Built in South Orleans around 1750, it was moved to the hill near Jonathan Young's home in 1839, where the Governor Prence motel now stands. In 1897, Captain Hunt, a retired sea captain, moved the mill to his estate in Hyannis Port. In 1983, it was moved back to Orleans.

Location

Jonathan Young Windmill
Town Cove Park, Route 6A
Orleans

Directions

From Eastham travel on Route 6 back to the Orleans rotary. Take the Orleans Center exit off the rotary to Route 6A. Cove Park and the Jonathan Young Windmill will be on the left, just after the set of lights.

Hours

Daily from late June through August, 11:00 am to 4:00 pm.

Contact Information

Orleans Historical Society
3 River Road and Main Street, Box 353
Orleans, MA 02653
Tel.: 508-240-1329

Reproduction windmill on Mill Pond Road, Orleans. Photo by the author.

Other Features

Not far from the Jonathan Young Windmill are a fine private reproduction windmill and two commercial reproductions. From Route 6A, turn left on Main Street toward Nauset Beach. At the second light, turn left on Tonset Road. After approximately one mile, turn right on Brick Hill Road. Take the second left onto Champlain Road, then the first right onto Mill Pond Road. Follow this down to 45 Mill Pond Road on the right. This reproduction at a private residence was built in the 1980s, based on the Eastham Windmill. Please view only from the road.

Return to Main Street in Orleans and turn left. In less than one mile you will see Nory's Gift and Card Shop on the left, in a windmill-style building built in the 1970s.

Leonore (Nory) and Brian Leonard, in front of Nory's Gift and Card Shop, Orleans, MA. Photo by the author.

Continuing toward Nauset Beach, there is another reproduction windmill at the Nauset Knoll Motor Lodge on the right, overlooking Nauset Beach.

The windmill at Nauset Knoll Motor Lodge, Orleans, MA. Photo by the author.

The Benjamin Godfrey Windmill at Chase Park, Chatham, MA. Photo by the author.

CHATHAM

Windmill viewing is rarely better anywhere in New England than in Chatham. The town has seen at least eleven working windmills since the 1700s, with as many as seven grinding at one time. Today, there is the splendid Benjamin Godfrey Windmill in Chase Park from the late 18th century, and "Sur Mer"– a mid-19th century gem in an unmatched setting. Several good reproductions attached to homes attest to the town's love of old mill design. A drive past these will reveal the unexplored back roads and harbors of this richly picturesque town.

The Windmills

The **Benjamin Godfrey Windmill** is owned by the town of Chatham and run by the Parks & Recreation Commission. Lou Springmeier, the mill keep, tells of the mill being built in 1797 on Mill Hill, off Stage Harbor Road. In 1956, it was donated to the town by Mr. and Mrs. Stuart Crocker and moved to its present location at Chase Park, off Cross Street.

Hours and Entrance Fees

Open July 1 through Labor Day, 10:00 a.m. to 3:00 p.m. Admission free.

Contact Information

Parks Department
Chatham Town Hall
Main Street
Chatham, MA 02633
Tel.: 508-945-5158

An early postcard of **Sur Mer,** Chatham, MA.

Located at 157 Shore Road, the **Sur Mer Windmill** is the most stunning privately owned windmill on the Cape.
A wildflower meadow slopes to its entrance, and picturesque Chatham Harbor lies beyond it in the immediate distance.

Known originally as the Seth Bearse Windmill, it was built about 1850 on the north side of Cockle Cove Road in South Chatham.

Access

"Sur Mer" is privately owned. View only from Shore Road, and please do not trespass.

Directions

From Orleans, take Route 28 to Chatham. At Old Harbor Road do not turn right to continue on Route 28, but cross Old Harbor Road, and follow the signs for the Chatham Lighthouse on Shore Road. Watch for Seaview Street on the right: The first house on the right has a splendid example of a Kendrick and Weintz reproduction windmill (private property).

A Kendrick and Weintz reproduction windmill, Seaview Street, Chatham, MA. Photo by the author.

Continue on Shore Road, past the Chatham Bars Inn on the right. Watch for the **Sur Mer Windmill**, privately owned, at 157 Shore Road.

Off Shore Road, take Main Street toward the center of Chatham. On the right, at 352 Main Street, will be The Dolphin of Chatham. This inn has a good Kendrick and Weintz reproduction windmill. Take a left at Cross Street, off Main Street, and watch for the **Benjamin Godfrey Windmill** at Chase Park on the left.

Continue on Cross Street and take a left on Stage Harbor Road, then turn left on Bridge Street. At the end of Bridge Street, turn right on Morris Island Road and take the causeway to Morris Island. Follow the winding road and turn right on Attucks Trail. At the end of Attucks Trail will be two Kendrick and Weintz reproduction windmills, a 16-foot mill on the left and a 13-foot one on the right. Both are on private property.

Reverse direction to leave Morris Island. At the end of Morris Island Road, where it joins with Shore Road, enter Windmill Lane. Directly ahead will be the delightful Robert Horne Reproduction Windmill, now owned by the Hallock family (private property).

Leaving Chatham on Route 28, there is Will's Mill on the left, at the corner of Cockle Cove Road. This reproduction mill was built in the early 1970s and over the years has housed a real estate agency and a restaurant. The mill was never completed, its arms never mounted.

Will's Mill reproduction, Route 28 and Cockle Cove Road, Chatham, MA,. Photo by the author.

Further Information

Chatham Historical Society
Old Atwood House Museum
347 Stage Harbor Road
Chatham, MA 02633
Tel.: 508-945-3678

Eldredge Public Library
564 Main Street
Chatham, MA 02633
Tel.: 508-945-5170

Day Four
NANTUCKET ISLAND

Nantucket's Old Mill is the last of five known windmills on the island. The stories of these mills are wreathed in myth and mystery: visions, ghosts, murder, ruin by lightning, arson, and deliberate, town-sponsored explosion—all give the writing of their histories the solidity of fog. (See Chapter Three for details.) A day trip from Hyannis to Nantucket Island is well worth the effort...in fact, for windmill followers it's virtually mandatory. The Old Mill of Nantucket is the only one in New England, perhaps in the country, with an apprenticeship program for windmillers. It is certainly the most dramatic working windmill in the northeast.

The Old Mill on Nantucket Island. Photo by the author.

The Windmill

According to historian Elizabeth Oldham, the **Old Mill** in Nantucket was built in 1746, and ground corn until 1892. In 1897, Caroline French purchased the mill at auction for $850 and donated it to the Nantucket Historical Association.

Location

The Old Mill
Mill Hill Park
Corner of South Prospect and South Mill Streets
Nantucket Island

Directions

Nantucket can be reached from Hyannis by two ferry services or by air:

The Steamship Authority offers high-speed (one hour) or regular (two hours and fifteen minutes) service from its dock on South Street, Hyannis. For schedule and information call 508-477-8600. Web Site: www.islandferry.com.

Hy-Line Cruises offers high-speed and regular ferry service from its Ocean Street dock in Hyannis. For schedule and information call 888-778-1132, or 508-778-2600. Web Site: www.hy-linecruises.com.

Nantucket Airlines offers flights from the Barnstable Municipal Airport in Hyannis. For schedule and information call 800-352-0714.
Web Site: www.flycapeair.com.

Hours and Entrance Fees

The Old Mill is open late May to mid-October from 10:00 am to 5:00 pm, Monday through Saturday. This and other Nantucket Historical Association properties may be visited using a History Ticket, $15 for adults, $8 for children, and $35 for families of one or two parents and their children. Purchase tickets in Nantucket at the Whaling Museum, 13 Broad Street, or the Hadwen House at 96 Main Street.

Further Information

Nantucket Historical Association
P.O. Box 1016
Nantucket, MA 02554
Tel.: 508-228-1894
Web Site: www.nha.org
E-mail: nhainfo@nha.org

Nantucket Historical Association Research Library
7 Fair Street
Box 1016
Nantucket, MA 02554
Tel.: 508-228-1655
E-mail: ralph@nha.org
Daily use fee: $5.00

Nantucket Antheneum
1 India Street
P.O. Box 808
Nantucket, MA 02554
Tel.: 508-228-1110
E-mail: antheneum@nantucket.net
Admission: Free

Day Five
RHODE ISLAND
EAST GREENWICH – JAMESTOWN – NEWPORT
MIDDLETOWN

Like all things in Rhode Island, a windmill tour of the state fits nicely in a very small space. Most of Rhode Island's remaining windmills are within easy reach of Newport on two bridge-connected islands—Conanicut and Aquidneck.

This Windmill Tour offers directions from Providence as a convenient starting point. We'll pass a windmill bought as a gift by Henry Wadsworth Longfellow, stop at the Jamestown Windmill on Conanicut Island, then enter Newport and contemplate whether its stone windmill was originally a Viking tower.

We'll also see Boyd's Windmill, where the famous Rhode Island Jonnycakes originated, and finish farther north on Aquidneck Island, where we'll meet the Duck Man at the Prescott Farm Windmill.

EAST GREENWICH

The Windmill

Poet Henry Wadsworth Longfellow had an enduring friendship with Professor George Washington Greene of East Greenwich. As an exceptionally generous sign of their friendship, Longfellow bought a home in 1866 as a gift for Greene and his wife. Four years later, Longfellow purchased an old windmill on the southeast corner of Division and East Streets and had it moved to the Greene house, which later became known as **Windmill Cottage**.

Location

Windmill Cottage
144 Division Street
East Greenwich

Directions

From Providence, take Route 95 toward Newport. At Exit 9 get on Route 4 South. Take the first exit off Route 4 onto Route 401 East to East Greenwich. This becomes Division Street. Look for Windmill Cottage at number 144 Division Street.

Access

Private home. View from street. Please do not trespass.

Further Information

Martha R. McPartland, *The History of East Greenwich, Rhode Island, 1677-1960.* East Greenwich Free Library Association, 1960.

The Jamestown Windmill, Jamestown, RI. Photo by the author.

JAMESTOWN

The Windmill

In 1787, Jamestown decided to build a town windmill and hire a miller to run it. For 109 years, the Jamestown Windmill ground cornmeal and animal feed until the big new rolling mills of the Midwest put it out of business. The **Jamestown Windmill** has been battered and broken by hurricanes, restored and restored again. A major restoration, at great cost, was completed in 1970. Thousands arrived for the celebration. Then a freak gale hit in March, 1974, snapping off all four vanes at the windshaft. The latest restoration occurred in 2002

Location

The Jamestown Windmill
North Road
Jamestown

Directions

From Windmill Cottage on Division Street in East Greenwich, take Division Street west toward Fairmont Drive. Merge onto Route 4 South via the ramp on left. Take a slight right onto Tower Hill Road/Route 1 South. Merge onto 138 East toward Jamestown/Newport Bridges. Cross the Jamestown Bridge and take exit toward Jamestown Center. Turn left onto North Road. Watch for the Jamestown Windmill on the right, just before the house at 382 North Road.

Hours

Open Mid-June to Mid-September, Saturdays and Sundays 1:00 pm to 4:00 pm, and by appointment.

Further Information

Jamestown Historical Society
Narragansett Avenue
Jamestown, RI 02835
Tel.: 401-423-0784

NEWPORT

The Windmill

The Newport Tower, Touro Park, Newport, RI. Photo by the author.

The most mysterious windmill in New England—and the most controversial—is the **Newport Tower** in Touro Park. For more than 250 years, scholars and poets have argued over the tower's origin. Was it a pagan temple, church, fort, gunpowder house, watchtower, lighthouse, windmill, or a combination of these? Was it built by the Druids, Phoenicians, Romans, Vikings, medieval explorers, Native American sachems, Rhode Island's first governor, a Portsmouth mason, or a Portuguese sailor? Only recent Carbon 14 dating has been able to finally settle the argument. (See Chapter Three for details.)

Location

Newport Tower
Touro Park
Corner of Bellevue Avenue and Mill Street
Newport

Directions

From the Jamestown Windmill, travel north on North Road. Merge onto Route 138 East. Cross the Newport Bridge and take the exit to Scenic Newport. Turn right onto 138A/238, Connell Highway. Continue to follow 138A/238 to Farewell Street. Take a slight right onto Thames Street, then turn left onto Mill Street. Continue until Touro Park appears on the right.

Access

The Newport Tower is an open stone structure, surrounded by a protective fence, and can be viewed from the fence, year-round, dawn to dusk.

Further Information

Newport Historical Society
82 Touro Street
Newport, RI 02840
Tel.: 401-846-0813
Fax: 401-846-1853
Web Site: www.newporthistorical.com

Redwood Library and Atheneum
50 Bellevue Avenue
Newport, RI 02840
Tel.: 401-847-0292
Web Site: www.redwood1747.org/tower/millmenu.htm

MIDDLETOWN

The Windmill

Boyd's Windmill began its long life in Portsmouth in 1810. In 1990, the Boyd family donated the mill to the Middletown Historical Society. The only known eight-vaned windmill in the U.S., it had been idle for nearly 50 years and its arms were gone. The "flaked" windmill was trucked from Mill Lane in Portsmouth to Paradise Park in Middletown on August 1, 1995. It took four more years to restore the mill to working condition, complete with its unusual array of eight sail-covered vanes.

Boyd's Windmill, Middletown, RI. Photo by the author.

Location

Boyd's Windmill
Middletown Historical Society
Paradise Valley Park
Middletown

Directions

From Touro Park in Newport, take Bellevue Avenue north. Turn right on Courthouse Street, then turn right on Broadway. Turn right on Miantomi Avenue, and stay straight to go onto Green End Ave. Boyd's Windmill will be on the right, at the end of Green End Avenue where it meets Paradise Avenue.

Hours and Further Information

Open July through September on Sundays, 2:00 pm to 4:00 pm, and for school programs.

Middletown Historical Society
P.O. Box 4196
Middletown, RI 02842
Tel.: 401-849-1870
E-mail: info@middletownhistory.org
Web Site: www.members.cox.net/mhs1875/index.htm

MIDDLETOWN

The Windmill

The striking **Prescott Farm Windmill**, a smock windmill with a dome cap, was built in Warren, Rhode Island, around 1812. In 1968, Doris Duke and her Newport Restoration Foundation purchased the windmill, and in 1970 moved it in sections five miles down West Main Road to Prescott Farm. Today, John Lingley, known to local children as "The Duck Man," offers tours of the buildings and, wearing his trademark hat trimmed out with duck feathers, cares for the farm's flock of ducks.

The Prescott Farm Windmill, Middletown, RI. Photo by the author.

Location

Prescott Farm
2009 West Main Road
Middletown

Directions

From Boyd's Windmill at the Middletown Historical Society, travel north on Paradise Avenue, which becomes Berkeley Avenue. Turn left on Wyatt Road, then turn right on Turner Road. Take a slight right onto Route 138, East Main Road, then left on Oliphant Lane. Turn right on Route 114, West Main Road, and Prescott Farm will be on the right, at 2009 West Main Road.

Hours and Entrance Fees

May through October, Monday through Friday, 10:00 am to 4:00 pm.
Guided tours: $3.00 adults, $1.00 children.
Tel.: 401-847-6230, or 401-849-7300

Further Information

The Newport Restoration Foundation
39 Mill Street
Newport, RI 02840
Tel.: 401-847-2071

Other Features

Buildings at Prescott Farm include the General Prescott House, circa 1730, and the John Earle House/Country Store, circa 1715.

Appendix Two: Millspeak: *A Compendium of Windmill Sayings, Aphorisms, and Terms*

Part One: Windmill Sayings and Aphorisms

And the sound of the millstone shall be heard in thee no more.

The Bible, New International Version, Revelation 18:22

As stout as a miller's waistcoat, that takes a thief by the throat every day.

German saying referring to the suspicion that millers cheated in their measurement of flour.

Back of the loaf is the snowy flour,
And back of the flour is the mill
And back of the mill is the wind and the shower,
And the sun and the Father's will.

Maltbie D. Babcock, 1858 – 1902

Come to a grinding halt.

If the millstones ground too close while the wind was dying, the mill would "come to a grinding halt."

Daily grind.

The repetitive nature of milling led to the concept of "the daily grind."

Dry-land sailors.

Another name for windmillers. Refers to the similar skills needed to sail a ship or run a windmill.

Fair to middling.

The quality of ground meal would be fair, middling, or fine. To be "fair to middling" is to be below one's best.

Flag on the mill, ship in the bay.

A saying from Sag Harbor, Long Island. Refers to signals conveyed by flags, and other means, from windmills to ships.

Folk made bread with the wind.

Said by the character John Ridd in R.D. Blackmore's *Lorna Doone*.

For if the flour be fresh and sound,
Who careth in what mill 'twas ground?

Henry Wadsworth Longfellow

Grist for one's mill.

A thought or idea to ponder, to "chew over."

Here lies an honest miller, and that is Strange.

Epitaph on an Essex, England, headstone.

I feel as stupid, from all you've said as if a mill wheel whirled in my head.

Goethe, from *Faust*, Act I

Less good from genius we may find
Than that from perseverance flowing
So have good grist and hand to grind
And keep the mill a-going.

Robert Burton, 1576 – 1640

Men grind and grind in the mill of a truism and nothing comes out but what was put in.

Ralph Waldo Emerson

The mill goes toiling slowly round with steady and solemn creak.

Eugene Field

The miller's hogs were always fat.

American saying.

Millery, millery, dusty soul,
How many sacks have you stole?

English nursery rhyme.

Milling around.

Revolving, and by extension wandering or meandering in one area.

Mills and wives are ever wanting.

English proverb.

A millstone and the human heart are driven ever round.
If they have nothing else to grind, they themselves must be ground.

Henry Wadsworth Longfellow

A millstone around one's neck.

"Things that cause people to sin are bound to come, but woe to that person through whom they come. It would be better for him to be thrown into the sea with a millstone tied around his neck than for him to cause one of these little ones to sin."

The Bible: New International Version, Luke 17:1-3

A millwright's sweat is strong enough to kill a snake.

A common saying, with "toad" sometimes substituted for "snake."

No mill, no meal; no will, no deal.

A Greek-Roman proverb.

No miller can enter heaven.

A saying from Normandy.

One who on earth has been a miller tells nought but lies afterward.

A saying from Normandy.

Put a miller, a weaver, and a tailor in a bag, and shake them. The first that comes out will be a thief.

English saying.

Put through the mill.

To be put through an ordeal, as corn is ground between stones.

Put your nose to the grindstone.

If millstones ground too hot, the flour would become cooked, emitting a burning smell. Occasionally, flour would burst into flames. The miller kept his "nose to the grindstone" to detect the temperature and condition of the meal.

Put your shoulder to the wheel.

When a miller had to turn a windmill into the wind, he "put his shoulder to the wheel" by pushing the wheel at the bottom of the mill's tailpiece. Some tailpieces had a yoke for the miller's shoulder. Some millers used a horse.

Rule of thumb.

To test the quality and grind of the flour, the miller would take a pinch of it between his thumb and finger. If too coarse, the flour would be ground again.

Run of the mill.

The ordinary, daily grind.

Safe as a thief in a mill.

Millers were sometimes considered to be thieves for the way they measured flour. A thief in a mill would presumably be among friends. See also, "What is bolder than a miller's neck-cloth...." and "As stout as a miller's waistcoat...."

The same old grind.

Similar to "the daily grind."

Show your metal.

Millstones often needed to be dressed (re-carved). When a miller hired an itinerant dresser, he could tell whether the man was experienced by noting the slivers of metal (thrown off from his carving tools) embedded in his hands. Variant of "show your mettle," "mettle" deriving from "metal."

Take your turn.

To "take your turn" is to be the next person to have corn or wheat ground by the turning of the millstones.

Though the mills of God grind slowly,
yet they grind exceeding small;
Though with patience he stands waiting,
with exactness grinds he all.

Friedrich von Logau, "Retribution," 1655

Three sheets to the wind.

If only three of the four arms are set with sails, a windmill is unbalanced and tipsy, like a drunk.

Tilting at windmills.

To foolishly go up against an imagined enemy. From Cervantes' *Don Quixote*. Quixote charges windmills which he believes are giants. "Tilting" refers to the thrusting of a lance in jousting.

Weep millstones.

To weep large, heavy tears.

What is bolder than a miller's neck-cloth which takes a thief by the throat every morning?

English saying.

Who comes first to the mill, first must grind.

Saxony proverb.

Who so cometh first to mille, first grynt.

From Chaucer's *Canterbury Tales*

The Windmill.

A tap dancing move that involves rotating the arms like a windmill.

You can never tell upon whose grain the miller's pig was fattened.

English proverb.

Your eyes are so sharpe that you cannot only look
Through a millstone, but cleane through the minde.

John Lyly, 1553 – 1601

Part Two: A Glossary of Terms

Ark

A storage bin for grain or meal.

Bails

Large iron tongs suspended from a crane, used to lift off the upper stone of a pair for dressing or to make other adjustments to the stones.

Balance Rynd

A curved iron bar that crosses the eye of the runner millstone, fitting into slots or pockets on either side. Also called a millstone bridge or crossbar. See also "Rynd."

Balance Weights

Used to balance the runner stone.

Beard

A decorative date board on Dutch windmills.

Bed Stone

The lower stationary millstone in a pair of millstones. Also called the "nether stone."

Bill

Cutting tool for dressing millstones.

Bins

Storage containers for grain, usually on the upper floor of a mill, from which grain would be fed into millstone hoppers. Also called "garners."

Bist

A cushion usually made of a partially filled sack of meal or bran, used as a cushion by a millstone dresser when dressing the millstones.

Blades

The arms or vanes of a windmill, attached to the windshaft.

Bolter

A machine used to sift flour into lots of different textures or degrees of fineness.

Bolting Cloth

Cloth of varying weave used to sift flour into lots according to texture and size. Sometimes made of silk, in which case referred to as "silks."

Brake

A band of wood or iron on the outer rim of the brake wheel, controlled by the brake lever, staff, or brake rope (known as the "gripe").

Bran

The outer coating of a grain of wheat, rye, barley, or corn. Oats and buck wheat have an outer coating that is a "hull."

Breast

The middle section of a millstone.

Breast Beam

The horizontal beam supporting the neck of the wind shaft.

Bridge Tree

A lever beam which carries the lower end of the spindle and thus bears the weight of the runner stone. It may be raised or lowered to alter the distance between the grinding surfaces of the millstones in order to produce a finer or coarser meal. See also "Tentering Staff."

Butterfly Furrow

The smallest of the four millstone furrows in one quarter of a millstone in quarter-dress. See also "Fly Furrow."

Canister

The iron casting on the poll end of the wind shaft to accommodate the blades and sails.

Cant Posts

The corner posts of a smock mill.

Cap

The top of a windmill, containing the mill's machinery. The variety of caps topping off windmills add to their distinctive regional styles. See also "Dome Cap," "Flemish Cap," "Kentish Cap," "Norfolk Cap," "Pent Roof Cap," and "Sussex Cap."

Cap Frame

The horizontal timber frame that supports the base of the cap.

Circular Furrow Dress Pattern

See "Sickle Dress Pattern."

Cracking

The act of cutting the fine grooves (cracks, drills, feathering, or stitching) along the lands of a millstone.

Cracks
The fine grooves cut into the face or lands of a millstone in the areas between the furrows.

Curb
The circular timber track on top of the body of a mill on which the cap turns. A live curb has wheels to make the turning of the cap easier; a dead curb does not.

Custom Mill
A relatively small milling operation that ground enough flour and meal to satisfy the needs of a local community. The miller was paid in kind, keeping a percentage of the ground meal for himself. The "miller's toll" was set by law and ranged from 10 to 20 percent. These small mills became known as "custom mills" because they ground from coarse to fine, according to the needs (custom) of the local people.

Damsel
A square or round shaft made of iron and/or wood. It terminates with a fork or crutch which, while rotating, taps the shoe, thus feeding grain into the millstones. Also called a "dandelion."

Dandelion
See "Damsel."

Dome Cap
A plain, dome-shaped windmill cap without the reverse curves of a Sussex (or Ogee) Cap. Common on the windmills of Rhode Island.

Drag Stick
A stick for unclogging grain from the eye of a millstone.

Drainage Mill
A windmill used for land drainage. Widely used in Holland.

Dress
The layout or pattern of furrows on a millstone.

Dresser
The person who chisels the millstone furrows.

Dressing

The process of cutting grooves (cracks or furrows) into the face of the mill stone, in order to provide the shearing action in grinding. Also the term for sharpening the existing dress. Also called "facing."

Dust Floor

The top floor of a smock mill.

Eye

The center hole in a millstone. In the runner stone the eye is always round. In a bed stone the center hole may be either round or square, depending upon the type of millstone-bearing housing used.

Face Gear (or Face Wheel)

A gear wheel with cogs mortised into its face, usually used in conjunction with a lantern pinion.

Facing

Another term for dressing.

Facing Hammer

A tool resembling a multiple-edged chisel used for dressing or facing a millstone.

Fang Staff

The brake lever.

Feed Shoe

A chute that guides grain from the hopper into the eye of the stone.

Flake

To flake a windmill is to disassemble it for the purpose of moving it to another site.

Flemish Cap

A cone-shaped windmill cap with a dormer, from which projected the wind shaft and the arms. It's a style older than the Kentish Cap, according to historian James Owens, and tops the windmill on the Eastham Common on Cape Cod.

Flour Dresser Machine

A machine for separating flour from the rest of the ground meal.

Fly Furrow

See "Butterfly Furrow."

Furrowing Stick

A wooden stick or straight edge used to mark out the line of the furrows used in dressing the millstones.

Furrows

The grooves cut into the grinding surface, or "lands" of the millstone.

Gallery

The stage around the body of some smock and tower mills. Used for access to the arms and sails.

Garners

See "Bins."

Grain Hopper

A hopper above the vat which holds the grain to be milled.

Grist

The material ground in a mill, e.g. corn, wheat, or rye.

Harp

A section of millstone face with a pattern of lands and grooves.

Hoop

See "Stone Case" or "Vat."

Hopper

An open-topped container tapered to feed grain into the millstones.

Hopper Ladder

See "Horse."

Horse

A wooden framework on top of the millstone case which holds the hopper, shoe, and damsel in position. Also called "a hopper ladder."

Journeyman Furrow

The second largest furrow in a quarter of a millstone in quarter dress pattern. It is parallel and immediately adjacent to the master furrow on one side and the apprentice furrow on the other side.

Kentish Cap

One of the oldest caps, the Kentish Cap was common on the early post mills. It's a round vault, like a tube cut in half, with flat ends. The style was common in Kent, England, and also in parts of Sussex and Essex.

Lands

The areas between the furrows on the grinding surface of a millstone. They are high parts of patterns on the surfaces of millstones.

Language of the Windmill

See "Saint Andrew's Cross" and "Saint George's Cross."

Lantern Pinion

A pinion gear consisting of round staves or rungs mortised between two discs, used either as a wallower, or as a millstone pinion.

Lighter Staff

See "Tentering Staff."

Loper

Another term for a runner stone.

Luff

The term for bringing the windmill into the wind.

Main Shaft

The vertical shaft from the wallower to the spur wheel.

Master Furrow

The largest furrow in a quarter of a millstone in quarter-dress, determining the boundary of the quarter.

Meal Floor

The floor below the millstones, where the freshly ground meal is received.

Meal Spout

A spout that conveys the meal from the millstones to the meal bin, or "bagger."

Middlings

The coarsest part of the wheat meal ground by a mill; the last product excepting the bran remaining after finer grades of flour are sifted out in the bolting process. A mediocre grade of flour, or the middle grade of flour. Also called "midds." See also "Sharps" and "Shorts."

Mill Bill

A chisel-ended tool used for dressing or sharpening the grinding surface of millstones. Also called "a mill chisel."

Mill Pick

See "Mill Bill."

Miller's Toll

The portion of ground meal retained by the miller as payment for his services. In the United States, usually 10 to 20 percent of the meal ground was the accepted toll. Local law governed how much the miller could take in tolling.

Millstone Bridge

See "Balance Rynd."

Millstones

A set of two millstones, consisting of an upper (or runner) millstone and the lower (or bed) stone. Also called a "run" (of stones).

Mortar and Pestle

A simple grinding apparatus in which a receptacle (mortar) is used to hold grain while it is crushed by a club-shaped implement (pestle).

Multure Bowl

The bowl used by the miller to measure his toll. See also "Miller's Toll."

Net Metering

If a modern home or business windmill produces more electricity than is needed at a specific time, utility companies will buy the extra electricity. In Net Metering, the meter turns backward while excess energy is supplied to the power company.

Nether Stone

The lower stone of a set of millstones. Also called the "bed stone."

Norfolk Cap

A particularly aerodynamic windmill cap, shaped like the bottom of a boat turned upside down. Like a boat, the sides are broad and taper at the ends. It was used in the Norfolk Broads of England, in Lancashire (in a larger version), and in the southwest of Denmark.

Outside Winder

A smock or tower mill whose cap is turned by a tail pole.

Overdrift Mill

A mill with the runner stone driven from above.

Paint Staff

A wooden, straight-edged staff used to test the surface level of a millstone. Also called a "proof staff."

Pair

A term used to describe the capacity of a mill, referring to the number of pairs of millstones installed; for example, a "two pair mill," a "three pair mill."

Pent Roof Cap

An early windmill cap, it is a simple peak, like the roof of a house. Many Cape Cod windmills have this type of cap.

Post Mill

The earliest type of windmill in America. The blades and all the machinery were contained in one box-like millhouse. The millhouse was held up by a single post, and the post was supported by trestles. The millhouse revolved on a tallowed wooden collar on the post so that the sails could face into the wind. The mill was turned when a man wheeled the tailpiece (a long pole sloping from the box to a wheel on the ground) into place.

Pottle

The portion of meal or flour paid to the miller. Also called the "toll."

Proof Staff

See "Paint Staff."

Quarter

A section of the surface of a millstone defined by master furrows, not necessarily one-fourth the surface area of a millstone.

Quarter Dress

A form of millstone dress using a series of straight furrows, the largest of which divide the surface of the millstone into regions called "quarters."

Quartering

Turning a windmill 90 degrees into the wind to halt it.

Quern

An early, simple form of rotary gristmill, consisting of a stationary lower bed stone and an upper runner stone usually rotated by hand with the aid of a stick.

Quill Stick

A flat piece of wood with a hole to accommodate a feather quill at one end. The other end contains a square hole that fits onto the millstone spindle. It is used to test the millstone spindle for balanced, upright running. Also called a "jack stick," "tram stick," or "trammel."

Raddle

A mixture of red oxide or lamp black powder and water, used on a paint staff. The mixture marks the high spots on a millstone to determine whether it is level. Also call "tiver."

Rap

The block on the shoe against which the damsel strikes (thereby causing the shoe to agitate), to ensure an even flow of grain from the hopper to the mill stones. This is usually made of hard wood while the shoe is made of soft pine. At times the rap is covered with a leather strap to dampen the sounds of the damsel striking against the shoe.

Rotor

On windmill power turbines, the rotors are the parts usually called "blades," or "arms" on traditional windmills.

Run

A pair of millstones. A term also used for the capacity of a mill. See also "Pair."

Runner Stone

The upper, moving millstone of a pair of millstones.

Rynd

A crossbar containing the bearing on which the upper runner stone of a pair of millstones rests and is balanced.

Sack Boy

A device for holding open sacks.

Sack Hoist

A method of hoisting sacks or barrels vertically in a mill using a gear-driven system or a windlass/barrel-hoist system. It was used to lift sacks from carts, wagons, boats, and the lower floors of the mill.

Saddle Stone Mill

An early, simple grinding apparatus in which meal is ground between a saddle-shaped stone and a rounded stone that is rolled over it.

Sail Stock

The assembly for sails, attached to the windshaft.

Saint Andrew's Cross

When the arms of a windmill were at rest in a balanced, diagonal cross, the position was called a "Saint Andrew's Cross."

Saint George's Cross

If the arms of a windmill were set in a vertical cross, the position was called a "Saint George's Cross." Said to have been used as a signal that "the miller is away." If there was a death in the community, the sails would be stopped just prior to the vertical position. Mourning could also be noted by removing some of the boards on the arms—the more boards removed, the closer the relationship of the miller to the deceased. If the miller himself died, many crossboards were removed and the arms turned slowly during the funeral.

Saltworks

Salt-producing equipment consisting of a windmill used to pump seawater into vats for evaporation.

Sharps

See "Middlings."

Shoe

A tapered trough vibrated to feed grain into the eye of the runner stone for grinding between the millstones.

Shorts

See "Middlings."

Shutdown Systems

In electric power windmills, if the windmill fails or the wind becomes too fierce, the system automatically turns the blades out of the wind and applies brakes.

Sickle Dress

A form of millstone dress using a series of semi-circular furrows. Also called "circular furrow dress."

Silks

See "Bolting Cloth."

Skirt

The outer edge of the grinding surface of a millstone.

Smock Mill

In the 17th century, a new style of tower mill was developed—the smock mill. These wooden towers sloped along eight sides, lending them the appearance of a countryman's linen smock. Made with large oak beams, pine floors and sheathing, and cedar shingles. The smock mill is the type of windmill most often seen in New England.

Spattle

A sliding shutter controlling the flow of grain from hopper to shoe.

Speed Control System

In electric power windmills, a system that shuts down the windmill if certain speeds are exceeded.

Spindle

The shaft on which the runner millstone rotates.

Stage

A projecting gallery around a tower or smock mill. Provides access to the sails and tail pole. Also known as the "reefing stage."

Stock

See "Sail Stock."

Stone Case

A circular wooden enclosure around a pair of millstones. Also called "casing," "hoop," "husk," "tun," or "vat."

Stone Dresser

A man whose profession is to re-sharpen, or "dress" millstones.

Storm Hatch

The access hatch over the neck bearing of the windshaft. Allows access from the mill to the blades and sails.

Strike

A straight-edge tool used to level off grain or meal in the toll dish. Also called a "strickle."

Sussex Cap

A round, somewhat onion-shaped windmill cap. It has a slight upward, reverse curve that ends in a finial knob. Also known as "ogee-shaped," it was also found in the East Midlands, the northeast of England, and the northeast of Denmark.

Swallow

The eye of the runner stone.

Sweepers

An attachment to the edge of the runner stone which sweeps the meal from the vat into chutes to the bins below. Also called the "tag."

Tag

See "Sweepers."

Tail Pole

A spar projecting from the cap of a mill to the ground. Used to rotate the cap and blades of the mill.

Tail-Winded

Describes a mill caught by a sudden change of wind, putting pressure on the wrong side of the sails.

Tentering

The process of adjusting the gap between the upper and lower millstones by raising or lowering the brayer and bridge tree.

Tentering Staff

A beam, or handle, connected to the bridge tree by the brayer, permitting the bridge tree to be raised or lowered and thus adjusting the gap between the upper and lower millstones. Also called a "lighter staff."

Thrift

A turned wooden handle that holds the bill.

Tidal Mill

A water mill harnessing energy from tidal water when it floods a salt marsh or the mouth of a river.

Tiver

A red ochre used for marking millstones. See also "Raddle."

Toll

A portion of meal or flour paid to the miller. Also called the "pottle."

Tower Mill

In the 16th century, the Flemish invented the tower mill, a technical advance over the post mill. With a simple shift in thinking, it was deduced that only the blades and cap of a windmill needed to turn, thereby allowing all the gears and the tons of millstones to remain stationary. There were two types of tower mills, the first being built of stone or brick and circular in shape. Often a wooden gallery or stage was built part way up the tower to allow the miller easy access to the sails. The second type was the smock mill.

Tracer

A wooden staff used to check true movement of the spindle.
See also "Quill Stick."

Tram Stick

See "Quill Stick."

Trammel

See "Quill Stick."

Tramp Iron

Loose bits of metal in the grain.

Tun

The removable wooden casing around millstones. Also called a "stone case" or "vat."

Urban Windmills

Small, stainless-steel electric power windmills mounted on roofs of businesses and homes in urban areas.

Vat

See "Tun" or "Stone Case."

Vertical Shaft

The long pole connecting the wind shaft to the runnerstone.

Wagon Wheel

The wheel attached to the bottom of the tail pole, used to rotate the mill.

Wallower

The gear wheel at the top of the vertical shaft. It connects to the brake wheel on the windshaft.

Warning Bell

A bell which rings when the grain content of the hopper gets too low.

Whips

Whips are the poles, bolted into the face of the stock, which hold the sail assemblies.

Wind Farms

When windmills are grouped together into a single wind power plant, they are known as "wind farms." They are constructed both on land and off-shore, where wind currents are more reliable.

Winding

The process of turning a mill to head into the wind.

Windshaft

The horizontal shaft carrying the blade and sail assembly (outside the mill) and the brake wheel (inside the mill).

Yoke

The wooden bars projecting downwards from the tail pole, against which the miller set his shoulders when winding (turning) the mill.

Notes

Introduction

1 John Reynolds, *Windmills & Watermills*. NY: Praeger Publishing, 1970, p.11.

2 Richard Bennett and John Elton, *History of Corn Milling, Vol. 1: Handstones, Slaves and Cattle Mills*. NY: Burt Franklin, 1898, p.192.

3 Laura Brooks, *Windmills*. NY: MetroBooks, 1999, p. 52.

4 John Reynolds, *Windmills & Watermills*. NY: Praeger Publishing, 1970, p.69.

5 Shebnah Rich, *Truro–Cape Cod*. Truro, Massachusetts, 1883, p.469.

6 Frederick Freeman, *The History of Cape Cod: The Annals of Barnstable County, Including the District of Mashpee, Volume I*. Boston: Printed for the Author, 1858, p.185.

7 Anne Barka, "The First Mill in English America," *Old Mill News*, July 1974, pp.4-5.

8 *Scientific American*, February 9, 1867.

9 John Tierney, "In New York, Change Is Traditional," *New York Times*, June 4, 2002, New York/Region section, p.4.

Chapter One

1 "Finding a poor widow, or dilapidated heir, having a share in some old tomb, the grave-digger, aware of the absence or death of the principal owner...purchases it for a trifie, seizes the whole...erases the family name, clears out the sacred relics which lie there, and then makes a trade by selling a berth for dead strangers in the city.... After the tombs have been filled up by the remains of strangers, their corpses have been carted out of town in the night...or buried in a hole dug at the bottom of the tomb, pounded down in one horrid, hideous mass, and covered over to make way for more death-money." Thomas Bridgman, *Epitaphs from Copp's Hill Burial Ground, Boston*. Boston: James Munroe and Co., 1851, pp.xx-xi.

2 *Second Report of the Record Commissioners of the City of Boston; Containing the Boston Records, 1634-1660, and the Book of Possessions*. Second Edition. Boston: Rockewell and Churchill, City Printers, 1881, pp.70-98.

3 Rex Wailes, "Notes on Some Windmills in New England," *Old-Time New England*, January 1931, pp.102-104.

4 David Ludlum, *The Country Journal New England Weather Book*. Boston: Houghton Miffiin Co., 1976, p.ix.

5 "Welcome to Wethersfield!" The Wethersfield Historical Society, http://www.wethhist.org/history.html.

6 Frederick Freeman, *The History of Cape Cod, Vol. 1*. Boston: Printed for the Author, 1858, p.197-198.

7 Elise Brett, "Eastham to Mark 300th Anniversary of Famed Windmill," *Cape Cod Times*, June 28, 1980, p.1.

8 Benjamin Franklin Wilbour, *Notes on Little Compton*. The Little Compton Historical Society, 1970, p.154.

Chapter Two

1 Sylvester Graham, *Treatise on Bread, and Bread Making*. Boston, MA: Light & Sterns, 1837.

2 See the Pond Lily Restorations web site: www.angelfire.com/journal/pondlilymill/

3 Douglas Heingartner, "Dutch Put New Spin on Windmills," *Sunday Republican*, Springfield, Massachusetts, January 26, 2003.

4 Henry David Thoreau, *Cape Cod*. Boston: Houghton Miffiin and Co., 1893, p. 38.

Chapter Three

1 *Fall River Herald News* (MA), February, 1942.

2 Interview with author, April 17, 2002.

3 Daniel Wing, *"Old Cape Cod Windmills,"* Library of Cape Cod History and Genealogy, No. 34. Yarmouth Port, MA: *Yarmouth Port Register Press*, 1924, p.7.

4 Lavinia Walsh, "The Oldest Wind Mill – A Cape Treasure," The *Cape Cod Magazine*, January 15, 1928, p.4.

5 Amanda B Harris, "A Windmill Pilgrimage," from *Sights Worth Seeing by Those Who Saw Them*. Boston: D. Lothrop & Co., 1886, p.40.

6 W.F. Kenney, "The Old Farris Grist-Mill," *Yarmouth Register*, Yarmouth, Mass., April 14, 1894.

7 Agnes Rothery, *The House by the Windmill*. Boston: Little, Brown, 1923, p.1.

8 Nathaniel Philbrick, "So Tottering a Tabernacle: The Wonder of the Old Mill," *Historic Nantucket*, Spring, 1996.

9 Clay Lancaster, "The Old Mill Reconsidered," *Historic Nantucket*, Spring, 1996.

10 Elizabeth Oldham, "The Old Mill: What We Know About It and What We Don't," *Historic Nantucket*, Spring, 1996, p.128.

11 Caroline Dusenberry Comstock, "My Narrow Escape," *Historic Nantucket*, Spring, 1996.

12 Helen Seager, "Portuguese Islanders and the Old Mill," *Historic Nantucket*, Spring, 2002, p.13.

13 Middletown Historical Society, *Wind Grist Mills on Aquidnick Island*. Middletown, Rhode Island, p.55.

14 *Fall River Herald News*, February, 1942.

15 Manuel Luciano Da Silva, "Portuguese Tower of Newport," *Portuguese Pilgrims & Dighton Rock*. Bristol, R.I., 1971, p.74.

16 Philip Ainsworth Means, *The Newport Tower*. NY: Henry Holt & Co., 1942, p.31.

Chapter Four

1 Frederick Freeman, *The History of Cape Cod, Vol. 2*. Boston: Printed for the Author, 1858, p.440-441.

2 Ibid, p.448.

3 *The Boston Globe,* "An Old Windmill," August 21, 1887.

4 Ibid.

5 Ibid.

6 Ibid.

7 Frederick Freeman, *The History of Cape Cod, Vol. 2*. Boston: Printed for the Author, 1858, p.246.

8 George F. Willison, *The Pilgrim Reader*. New York: Doubleday & Co., 1953, p.142.

9 *Barnstable Town Records*, January 19, 1687, Vol. 1, p.36.

10 Donald Wood, *Cape Cod: A Guide*. Boston: Little, Brown, pp.102-103.

11 From a copy in the William Brewster Nickerson Room, Cape Cod Community College.

12 Nancy Thacher Reid, *Dennis, Cape Cod: From Firstcomers to Newcomers, 1639-1993*. Dennis Historical Society, 1996, p.376.

13 Henry David Thoreau, *Cape Cod*. Boston: Houghton Miffiin, pp.263-264.

14 Eleanor Mayhew, *"Martha's Vineyard,"* Dukes County Historical Society, Inc., 1956, p.57.

15 Robert M. Downie, "Windmills – Nearly 200 Years of Nonstop Motion on Block Island," *The Block Island Times Digital Edition*, www.blockislandtimes.com, August 24, 2000, p.1.

16 Ibid, p.3.

17 Ibid, p.4.

18 Ethel Colt Ritchie, *Block Island Lore & Legends*. Published by the Author, 1980, p.57.

19 Robert M. Downie, p.5.

20 Walter Hill Crockett, *Vermont: The Green Mountain State*. NY: Century History Co., 1921, p.129.

21 Ibid, p.128-129.

Chapter Five

1 *Sunday Cape Cod Standard-Times,* "Mill Heads for History," May 19, 1971, p.28.

2 Richard Fox, "Old Dutch Mill Move Had Just One Hitch," *Sunday Cape Cod Standard-Times*, June 20, 1971, p.18.

3 Paul L. Chesbro, *Osterville*. Taunton, MA: William Sullwold Publishing, 1988, pp.18-19.

4 Sarah Korjeff, "Tilting at One Inviting Windmill," *Cape Cod Voice*, May 10, 2001.

Chapter Six

1 Published in *Criticism*, Summer, 1997.

2 http://www.storiesbypat.homestead.com.

Chapter Seven

1 Miguel de Cervantes, *The Ingenious Gentleman Don Quixote de la Mancha*. NY: Viking Press, 1949, pp.62-64.

2 Hans Bierdermann, *The Dictionary of Symbolism*. NY: Facts On File, 1992, pp.382-383.

3 Henry David Thoreau, *The Journal of Henry D. Thoreau, Vol.1*. NY: Dover Publications, 1962, August 26, 1851, p.251.

4 Henry David Thoreau, *Collected Poems of Henry Thoreau*. Baltimore: The Johns Hopkins Univ. Pr., 1970, p.24

5 W.H. Auden, *Collected Shorter Poems, 1927-1957*. NY: Random House, 1967, p.201.

6 Windmill historian John Reynolds has suggested that medieval millers may even have extended to the windmill the mystical ideal of perpetual motion.

7 Charles Perrault, *Puss in Boots*. NY: Farrar Straus Giroux, 1990, p.16.

8 Leo Tolstoy, *Anna Karenina*. NY: Modern Library, Random House, 1965, p.553.

9 Charles Dickens, *Great Expectations*. NY: W.W. Norton, 1999, p.236.

10 Poe, Edgar Allan, *The Narrative of Arthur Gordon Pym of Nantucket*. Boston: David R. Godine, 1973, p.66.

11 Herman Melville, *Mardi and a Voyage Thither*. NY: Harper & Brothers, 1849, p.227.

12 R.D. Blackmore, *Lorna Doone*. London: S. Low, Marston, 1873, p.61.

13 Robert Louis Stevenson, "To Nelly Sanchez," *Collected Poems*. London: Rupert Hart-Davis, 1950, p.324.

14 John Greenelaf Whittier, "The King's Missive," *The Complete Poetical Works*. Boston: Houghton, Miffiin and Co., 1892, p.419.

15 The Middletown Historical Society, *Wind Grist Mills on Aquidneck Island*. Middletown, Rhode Island, p.45.

16 Isaac M. Small, *Cape Cod Stories*. New Bedford, MA: Reynolds Printing, 1934, pp.49-50.

17 Kingsley Amis, "Ode to the East-North-East-by-East Wind," *Collected Poems*. London: Hutchinson, 1979, p.54.

18 Cate Marvin, "A Windmill Makes a Statement," *New England Review*, Spring, 2001, p.94.

19 Sylvia Plath, *The Unabridged Journals of Sylvia Plath*. NY: Anchor Books, 2000, p.233.

20 Anthony Breznican, "They Might Be Giants Celebrate 20-Year-Old Gig," *Sunday Republican*, August 11, 2002, Section D, p.2.

21 http://artists3.iuma.com/IUMA/Bands/Windmills/

Chapter Eight

1 "Turbine Pioneer, Charles F. Brush," *Scientific American*, December 20, 1890.

2 Robert M. Downie, "Windmills – Nearly 200 Years of Nonstop Motion on Block Island," *Block Island Times, Digital Edition*, August 24, 2000, www.blockislandtimes.com.

3 "Survey Shows Windmills Unwelcome," *Block Island Times, Digital Edition*, January 27, 2001, www.blockislandtimes.com.

4 Elizabeth Mehren, "Battle Lines Form Over Wind Power," *Los Angeles Times*, reprinted in the *Sunday Republican* (Springfield, Massachusetts), August 11, 2002, Section B, p.5.

5 Ibid.

6 Ibid.

7 Doreen Leggett, "Winds of Change," *The Cape Codder*, September 13, 2002, p.6.

8 Jack Coleman, "Barnstable Joins Push to Harness Wind Power," *Cape Cod Times*, May 4, 2002, p.1.

9 See the U.S. Department of Energy Wind Energy Program web site, www.eren.doe.gov/wind.

10 David Pitt, "Largest Wind Farm Planned for Iowa," *Sunday Republican* (Springfield, Massachusetts), April 12, 2003, Section E, p.12.

Bibliography

Articles

Barka, Anne. "The First Mill in English America," *Old Mill News*, July 1974, pp.4-5.

Coggeshall, Charles P. "Some Old Rhode Island Grist Mills," *Bulletin of the Newport Historical Society*, No. 39, January 1922.

Hamilton, E.P. "Some Windmills of Cape Cod," *The Newcomen Society Transactions, Vol. V*, 1924-1925, London, 1925.

Harris, Amanda B. "A Windmill Pilgrimage," *from Sights Worth Seeing by Those Who Saw Them*, Boston: D. Lothrop & Co., 1886.

Koehler, Margaret. "Cape Cod's Wandering Windmills," *Cape Cod Compass*, East Sandwich, MA, 1976, pp. 14-21, 52-55.

Shelton, R.H. "Windmills, Picturesque & Historic," *Journal of the Franklin Institute*, 1919.

Turner, Laurence J. "Preserved Heritage: Some Remaining Windmills of the U.S.A.," *Industrial Archeology*, 17 (2-4): pp.181-186, 1983.

Wailes, Rex. "Notes on Some Windmills in New England," *Old-Time New England, the Bulletin of the Society for the Preservation of New England Antiquities, Vol.XXI, No. 3*, January 1931, pp.98-128.

"Windmills in New York City," *Scientific American*, Feb. 9, 1867, p. 89.

Pamphlets

Wing, Daniel. "Old Cape Cod Windmills," *Library of Cape Cod History and Genealogy, No. 34*. Yarmouth Port Register Press, 1924.

Books

Baker, Lindsay. *A Field Guide to American Windmills*. Norman, OK: University of Oklahoma Press, 1985.

Balliett, Blue. *Nantucket Hauntings*. Camden, Maine: Down East Books, 1990.

Beedell, Suzanne. *Windmills*. NY: Scribners, 1975.

Bennett, Richard and John Elton. *History of Corn Milling, Vol. 1: Handstones, Slaves and Cattle Mills*. NY: Burt Franklin, 1898.

Berger, Josef. *Cape Cod Pilot*. Cambridge, MA: M.I.T. Press, 1969.

Bierdermann, Hans. *The Dictionary of Symbolism*. NY: Facts On File, 1992.

Brangwyn, Frank & Hayter Preston. *Windmills*. NY: Dodd, Mead and Co., 1923.

Brooks, Laura. *Windmills*. NY: MetroBooks, 1999.

Brown, Joseph E. and Anne Ensign Brown. *Harness the Wind: The Story of Windmills*. NY: Dodd, Mead and Co., 1977.

Burrows, Fredrika A. *Windmills on Cape Cod & the Islands*. Taunton, MA: William S. Sullwold Publishing, 1978.

Dennis, Landt. *Catch the Wind: A Book of Windmills and Windpower*. NY: Four Winds Press, 1976.

Freeman, Frederick. *The History of Cape Cod...In Two Volumes*. Boston: Printed for the Author, 1858.

Freese, Stanley. *Windmills & Millwrighting*. NY: Cambridge University Press, 1957.

Graham, Sylvester. *Treatise on Bread, and Bread Making*. Boston, MA: Light & Sterns, 1837.

Hopkins, Robert Thurston. *Old Watermills and Windmills*. London: P. Allan & Co., Ltd., 1930.

Larkin, David. *Mill: The History and Future of Naturally Powered Buildings*. NY: Universe Pub., 2000.

Lincoln, Joseph. *Cape Cod Yesterdays*. Boston, MA: Little, Brown and Co., 1935.

Marks, William. *The History of Wind-Power on Martha's Vineyard*. Martha's Vineyard, MA: The National Association of Wind-Power Resources, Inc., 1981.

Middletown Historical Society. *Wind Grist Mills on Aquidnick Island*. RI: Middletown Historical Society, 1992.

Mitchell, Edwin Valentine. *It's an Old Cape Cod Custom*. NY: Vanguard Press, 1949.

Preston, Hayter and Frank Brangwyn. *Windmills*. NY: Dodd, Mead and Co., 1923.

Quinn, William P. *The Saltworks of Historic Cape Cod*. Orleans, MA: Parnassus Imprints, 1993.

Reynolds, John. *Windmills & Watermills*. NY: Praeger Publishing, 1970.

Skilton, C.P. *British Windmills and Watermills*. London: Collins, 1947.

Stokhuyzen, Frederick. *The Dutch Windmill*. London: University Books, 1963.

Thoreau, Henry David. *Cape Cod*. Boston: Houghton Mifflin and Co., 1893.

Thoreau, Henry David. *Collected Poems of Henry Thoreau*. Ed. by Carl Bode. Baltimore: The Johns Hopkins Univ. Pr., 1970.

Thoreau, Henry David. *The Journal of Henry D. Thoreau*. NY: Dover Publications, 1962.

Torrey, Volta. *Wind-Catchers: American Windmills of Yesterday and Tomorrow*. Brattleboro, VT: Stephen Greene Press, 1976.

Wood, Donald. *Cape Cod: A Guide*. Boston: Little, Brown and Company, 1973.

Zimiles, Martha and Murray. *Early American Mills*. NY: Clarkson N. Potter, 1973.

Fiction

Blackmore, R. D. *Lorna Doone*. London: S. Low, Marston, 1873.

de Cervantes, Miguel. *The Ingenious Gentleman Don Quixote de la Mancha*, trans. by Samuel Putnam. NY: Viking Press, 1949.

Dickens, Charles. *Great Expectations*. NY: W.W. Norton, 1999.

Feil, Hila. *The Windmill Summer*. NY: Harper & Row, 1972.

Lincoln, Joseph. *Shavings: A Novel*. NY: Appleton & Co., 1918.

Martin, Vicky. *The Windmill Years*. NY: St. Martin's Press, 1978.

Melville, Herman. *Mardi and a Voyage Thither*. NY: Harper & Brothers, 1849.

Perrault, Charles. *Puss in Boots*, trans. by Malcolm Arthur. NY: Farrar Straus Giroux, 1990.

Poe, Edgar Allan. *The Narrative of Arthur Gordon Pym of Nantucket*. Boston: David R. Godine, 1973.

Rothery, Agnes. *The House by the Windmill*. NY: Little, Brown and Co., 1923.

Sheldon, Sidney. *The Windmills of the Gods*. NY: William Morrow and Co., 1987.

Tolstoy, Leo. *Anna Karenina*, trans. by Constance Garnett. NY: Modern Library, Random House, 1965.

Poetry

Amis, Kingsley. "Ode to the East-North-East-by-East Wind," *Collected Poems: 1944-1979*. London: Hutchinson, 1979.

Longfellow, Henry Wadsworth. "The Windmill," *The Complete Poetical Works of Henry Wadsworth Longfellow*. Boston: Houghton Mifflin Co., 1922.

Marvin, Cate. "A Windmill Makes a Statement," *New England Review*, 22:2, Spring, 2001.

Small, Isaac, "The Old Mill," *Cape Cod Stories*. New Bedford, MA: Reynolds Printing, 1934.

Stevenson, Robert Louis, "To Nelly Sanchez," *Collected Poems of Robert Louis Stevenson*. London: Rupert Hart-Davis, 1950.

Whittier, John Greenleaf, "The King's Missive," *The Complete Poetical Works of John Greenleaf Whittier*. Boston: Houghton, Mifflin and Co., 1892.

Songs

Balin, Marty. "Miracles," Diamondback Music Co., 1975.

Bergman, Alan, and Michel Legrand. "The Windmills of Your Mind," *Windmills of Your Mind* (album), United Artists, UAS 6715. Published by EMI U Catalog, Inc., 1970.

Puff Daddy. "Is This the End? (Part Two)," *Forever* (cd), released October 24, 1999.

Films

Army of Darkness, 1992. Director: Sam Raimi.

Foreign Correspondent, 1940. Director: Alfred Hitchcock.

Frankenstein, 1931. Director: James Whale.

Man of La Mancha, 1972. Director: Arthur Hiller.

Moulin Rouge, 1952. Director: John Huston.

Moulin Rouge, 2001. Director: Baz Luhrmann.

Organizations

European Wind Energy Association
Wind Directions, the Official Magazine
26 Rue du Trone
B-1000 Brussels
Belgium

The International Molinological Society
Michael Harverson
125 Parkside Dr.
Watford, Hertfordshire WD1 3BA
England
Web Site: http://www.tims.geo.tudelft.nl/

The National Wind Coordinating Committee
1255 23rd St., N.W.
Washington, D.C. 20037
Web Site: www.nationalwind.org

Society for the Preservation of Old Mills (SPOOM)
Dick Sullin, Treasurer
111 South Main St.
Rockford, MI 49341
Web Site: www.spoom.org

Society for the Preservation of Old Mills (SPOOM)
Northeast Chapter
Thomas F. Glick
132 Brook St.
Holliston, MA 01746
Journal: *Old Mill News*.
Web Site: www.bu.edu/medieval/spoom.htm

The Windmill Study Unit of the American Topical Association
Quarterly Bulletin
6125 Teagarden Circle
Dayton, OH 45449-3013

Web Sites

American Wind Energy Association
www.awea.org

British Wind Energy Association
www.bwea.com/

Canadian Wind Energy Association
www.canwea.ca/

The Danish Wind Industry Association
International information about wind energy.
www.windpower.org

The International Molinological Society
http://www.tims.geo.tudelft.nl/

The National Wind Coordinating Committee
www.nationalwind.org

The National Wind Technology Center
www.nrel.gov/wind/

Pond Lily Restorations
www.angelfire.com/journal/pondlilymill/

Society for the Preservation of Old Mills
www.spoom.org

U.S. Department of Energy Wind Energy Program
www.eren.doe.gov/wind/

Windmill World
www.windmillworld.com/

Windmills of Cape Cod
www.windmillworld.com/world/capecod

Windpower Monthly News Magazine
www.wpm.co.nz

Windstats Newsletter
Danish newsletter of wind energy.
www.windstats.com

Index

Acknowledgments

Karen Banta, Adam Gamble, James E. Owens, William Marks, Stuard Derrick, Barbara Townson Weller, Leslie Weller, Peter and Peggy Hallock, Jane Sugden, Dawn Carlson, John and June McCahill, Joseph Colliano and Wills Hastings, Richard and Edith Gallant, Hope Morrell (Cape Cod National Seashore Archives, MA), Mary Sicchio and Charlotte Price (William Brewster Nickerson Room, Cape Cod Community College, MA), Mary Ann Gray (Archivist, Chatham Historical Society, MA), Frank Kenney, Bill Putman, Elizabeth, David, and Emmy Roache, Samuel Barber, Steve Conner, Mary D. Blake (Mrs. Turner Blake), Herb Feeney, John Flender, Jane LeGrow and Candy Massard (Heritage Plantation, Sandwich, MA), Louis Springmeier (Godfrey Windmill, Chatham, MA), Andrew Shrake, Barbara Ryan (Director, Historical Society of Old Yarmouth, MA), Dan Lyons and John Falvey (Cape Cod Airport, Marstons Mills, MA), Marilyn Scofield (Eastham Historical Society, MA), Louis Nickinello, Tom Fedele, Bill White, Tom Gerhardt (Aptucxet Trading Post Museum, Bourne, MA), Falmouth Historical Society (MA), Cecil Barron Jensen (Nantucket Historical Association, MA), Patrick Prugh, Amanda Nicholas, Emily Chiswick-Patterson (Old Mill, Nantucket Historical Association, MA), Truro Historical Society (MA), Jonathan Fitch, Jonathan Allard, and Delores Lyons (Princeton Municipal Light Department, MA), Stanley Grossman, Mary Bellagamba, and William Miller (Middletown Historical Society, RI), M. Joan Youngken and Bertram Lippincott III (Newport Historical Society, RI), John Lingley (Prescott Farm Windmill, Middletown, RI), Rhode Island State Archives (Providence), Rhode Island Historical Society Library (Providence), Laura Anderson (Little Compton Historical Society, RI), Donald Malcarne (Essex Town Historian, CT), Sally Foote (Castine Historical Society, ME), Wanda Wood (Castine Realty, ME), Harley Lee (Endless Energy Corporation, Yarmouth, ME), Martha Staskus and Dorothy Schnure (Green Mountain Power Corporation, Colchester, VT), Art Miller (Searsburg Wind Power Facility, VT).

For Bonus Material on
Windmills of New England
and Dan Lombardo:
www.oncapepublications.com

Other Books from On Cape Publications

Quabbin: A History & Explorer's Guide by Michael Tougias

The Blizzard of '78 by Michael Tougias

Haunted Inns of New England by Mark Jasper

1880 Atlas of Barnstable Country: Cape Cod's Earliest Atlas edited by Adam Gamble

In the Footsteps of Thoreau by Adam Gamble

Cape Cod (Audio) by Henry David Thoreau

Sea Stories of Cape Cod & the Islands by Admont G. Clark

The Cape Cod Christmas Cookbook by Mark Jasper

Haunted Cape Cod & the Islands by Mark Jasper

Coloring Cape Cod, Martha's Vineyard & Nantucket by James Owens

Cape Cod, Martha's Vineyard & Nantucket, the Geologic Story by Robert Oldale

Cape Cod: Visions of a Landscape photos and text by Brian Smestad

Walking the Shores of Cape Cod by Elliott Carr

Baseball by the Beach: A History of America's National Pastime on Cape Cod by Christopher Price

A Guide to Nature on Cape Cod & the Islands edited by Greg O'Brien

Cape Cod Light: The Lighthouse at Dangerfield by Paul Giambarba

Penikese Island of Hope by Thomas Buckley

Howie Schneider Unshucked: A Cartoon Collection about the Cape, the Country and Life Itself by Howie Schneider

www.oncapepublications.com

BOYD'S
MEAL